Mekdes Girma

Concepts for Short Range Millimeter-wave Miniaturized Radar Systems with Built-in Self-Test

Karlsruher Forschungsberichte
aus dem Institut für Hochfrequenztechnik und Elektronik

Herausgeber: Prof. Dr.-Ing. Thomas Zwick

Band 92

Eine Übersicht aller bisher in dieser Schriftenreihe erschienenen Bände finden Sie am Ende des Buches.

Concepts for Short Range Millimeter-wave Miniaturized Radar Systems with Built-in Self-Test

by
Mekdes Girma

KIT Scientific Publishing

Karlsruher Institut für Technologie
Institut für Hochfrequenztechnik und Elektronik

Concepts for Short Range Millimeter-wave Miniaturized Radar Systems with Built-in Self-Test

Zur Erlangung des akademischen Grades eines Doktor-Ingenieurs von der KIT-Fakultät für Elektrotechnik und Informationstechnik des Karlsruher Instituts für Technologie (KIT) genehmigte Dissertation

von M.Sc. Mekdes Gebresilassie Girma

Tag der mündlichen Prüfung: 7. Dezember 2018
Hauptreferent: Prof. Dr.-Ing. Thomas Zwick
Korreferent: Prof. Dr.-Ing. Christian Waldschmidt

Impressum

KIT Scientific Publishing

Karlsruher Institut für Technologie (KIT)
KIT Scientific Publishing
Straße am Forum 2
D-76131 Karlsruhe

www.ksp.kit.edu

Print on Demand 2019 – Gedruckt auf FSC-zertifiziertem Papier

ISSN 1868-4696
ISBN 978-3-7315-0938-7
DOI 10.5445/KSP/1000095711

Preface of the Editor

Radars are no longer limited to space and defense applications. In fact, radars are being used in several areas of our everyday lives, without us being explicitly aware of their presence. Especially, the motion detectors, e.g., door openers and security alarms, are nearly ubiquitous. Thanks to the tremendous advancement of the semiconductor technologies, that has taken place in the last 15 years, radar systems are now being extensively used in automotive sensors as well. The resolution provided by a radar system is lower than that of an optical system, but a radar system, in contrast to an optical system, can deliver reliable measurement values in harsh weather conditions as well. This property is one of the main reasons, why more and more radars are being deployed in the industrial arena as well.

An important industrial application at present is contactless level measurement of liquids but many other applications will be conquered once the high potential for miniaturization and low cost per unit is lapped. One of the major cost drivers of high volume millimeter wave radar sensors is the testing. This is exactly where the work of Dr. Mekdes Girma comes into play. In her dissertation, Ms. Girma has worked upon the important scientific principles of near field radars with built-in self-test functionality. She has built several miniaturized radar systems and thereafter investigated their near field measurement capability, together with the built-in self-test functionality, in both theory and practice.

Ms. Girma has for the first time, built a heterodyne radar system and investigated its measurement accuracy in the extremely near-field region. Based on this investigation, she has shown that the "‘near-field peak"’ is highly stable over time in the new system and hence it can be calibrated out. Further, she has developed a novel radar concept with an integrated auto-calibration function and built a demonstrator for the same. With this, she has shown that the non-ideal characteristics and systematic errors can be calibrated out. This concept can also be used for observing the functionality of radars during their operation. The work of Dr. Mekdes Girma thus represents an important innovation to the state of the art. I am sure that her realized radar concepts and sensors will draw much attention worldwide. For Dr. Girma, with her very good technical skills, I wish her further success in her future career.

Prof. Dr.-Ing. Thomas Zwick
- Institute Director -

Concepts for Short Range Millimeter-wave Miniaturized Radar Systems with Built-in Self-Test

Zur Erlangung des akademischen Grades eines

DOKTOR-INGENIEURS

von der Fakultät für
Elektrotechnik und Informationstechnik
des Karlsruher Instituts für Technologie (KIT)

genehmigte

DISSERTATION

von

M.Sc. Mekdes Gebresilassie Girma

Tag der mündlichen Prüfung:	07.12.2018
Hauptreferent:	Prof. Dr.-Ing. Thomas Zwick
Korreferent:	Prof. Dr.-Ing. Christian Waldschmidt

Vorwort

Die vorliegende Arbeit entstand während meiner Zeit als Doktorandin bei der zentralen Forschung der Robert Bosch GmbH in Kooperation mit dem Institut für Hochfrequenztechnik und Elektronik (IHE) des Karlsruher Instituts für Technologie (KIT).

Zunächst geht mein Dank an Herrn Prof. Dr.-Ing. Thomas Zwick, Leiter des Instituts für Hochfrequenztechnik und Elektronik, für die wissenschaftliche Betreuung der Arbeit sowie die Wertvollen Anregungen und seine stets hilfreiche Unterstützung. Weiterhin möchte ich mich bei Herrn Prof. Dr.-Ing. Christian Waldschmidt, Leiter des Instituts für Mikrowellentechnik, für die Übernahme des Korreferats und der damit verbundenen Mühen bedanken.

Mein Dank geht an die Firma Robert Bosch GmbH, die es mir durch die Anstellung als Doktorandin ermöglichte, auf diesem Themengebiet zu forsche und somit die vorliegende Dissertation anzufertigen. Mein herzlicher Dank geht an meinen Betreuer Dr.-Ing. Jürgen Hasch für die fachliche Betreuung, Mentoring, sowie den Freiraum bei der Ausgestaltung der Arbeit. Seine fachliche Kompetenz diente als Vorbild und hat zu meiner Weiterentwicklung wesentlich beigetragen. Ein ganz besonderer Dank gilt meinem Kollegen Dr.-Ing Markus Gonser, und Andreas Frischen mit deren Hilfe ich zahlreiche technische Probleme lösen konnte. Ebenfalls geht mein besonderer Dank an gesamten Forschungsgruppe CR/ARE1 der Robert Bosch GmbH, für die ausgezeichnete Arbeitsatmosphäre und für die Unterstützung bei technischen sowie nichttechnischen Angelegenheiten.

Ein großer Teil der in dieser Arbeit beschriebenen Forschungsergebnisse wurde innerhalb des EU-Projekts SUCCESS erzielt. Aus diesem Grund möchte ich mich bei allen Projektpartnern der beteiligten Unternehmen und Institutionen für die stets zielführende und angenehme Zusammenarbeit bedanken. Diese betrifft insbesondere Prof. Sorin Voinigescu und Dr. Ioannis Sarkas von der Universität Toronto, Dr.-Ing Stefan Beer von KIT und Dr.

Yaoming Sun von IHP, die viele dieser Ideen erprobten und per Chip realisieren mussten.

Nicht zuletzt möchte ich mich bei meinen Freunden und bei meiner Familie für die beständige Unterstützung und den stetigen Zusammenhalt bedanken. Ich möchte mich ganz herzlich bei meinen Freunden Dr.-Ing. Berhane Gebresilassie, Emebet Legesse und Elazer Berhane bedanken, die mich stets unterstützt haben. Insbesondere danke ich meinen Eltern, Kebebush Haile und Gebresilassie Girma, sowie meinen Partner Thomas für Ihre liebevolle Unterstützung auf dem Weg zu meiner Promotion. Diese Dissertation möchte ich besonders meinem Sohn Michael widmen. Er hatte sicherlich seinen besonderen Beitrag bei Ihrer Entstehung geleistet.

Leonberg, im Oktober 2018
Mekdes G. Girma

Abstract

In the last decade, microwave and millimeter-wave (mm-wave) radar systems have gained importance in automotive, consumer and security applications. Due to an increasing maturity and availability of circuits and components, these systems are getting more compact while being less expensive. This recent interest is largely due to the advantages mm-wave radars offer over other range measuring sensors, as their performance is less affected by dust, fog, rain or snow and ambient lighting conditions. In the area of consumer radar, high integration and miniaturization of the whole system is the most important requirement to address. In this context, due to the short wave length, mm-wave radar systems operating above 100 GHz play a major role in-terms of high integration of the integrated circuit (IC) as well as the antenna in a housing. However, a major challenge of the high integration is, testing of the integrated mm-wave ICs become more difficult due to the absence of probe pads to access the components for contact based testing. To overcome this issue, a Built-in-Self-Test (BIST) approach may prove to be the most efficient test scheme.

This PhD work presents a series of single chip, mm-wave radar sensor systems for short range applications which are operating at 122 GHz. First, trends and emerging applications in millimeter-wave band are over-viewed and discussed. Typical system architectures of mm-wave radar sensors, such as homodyne versus heterodyne, monostatic versus bistatic, antenna and packaging concepts with all related benefits and drawbacks with focus on short range application are briefly discussed and investigated.

In the first part, a highly integrated, single chip, dual channel radar sensor system is investigated for the first time. The designing and implementation processes, starting from the system level design considerations and characterization of the individual components to final implementation of the proposed architecture are described briefly. Dual channel on-chip antennas at 122 GHz are employed and the functionality is demonstrated, which shows a distance and angle detection capability of the radar sensor.

In the second part, it is shown, how heterodyne radar systems with an intermediate frequency (IF) can overcome some of the commonly encountered problems with homodyne systems, i.e., DC-offsets resulting from hardware imperfection and short-range leakage and reflections. The measurement results show that better performance can be achieved in the heterodyne configuration, compared to similar work published previously. Further more, a case-study on the impact of temperature and random noises on the sensor performance, and a dynamic DC-offset issues is investigated.

In the final part of this work an embedded auto-calibration methodology and a novel BIST for radar systems is developed and investigated, for the first time. As testing of modern complex electronic systems is a very expensive procedure, special structures for simplifying this process can be inserted into the system during the design phase. A two channel heterodyne radar architecture with built-in-self-test and calibration features is proposed. A methodology for auto embedded calibration and built-in testing concept is briefly described. To verify the proposed method, a single chip sensor is realized at 122 GHz frequency and the concept is verified by using a prototype measurements. In addition, a complex S-parameter measurement of a mm-wave antenna is shown for the first time.

Zusammenfassung

Seit dem letzten Jahrzehnt werden zunehmend Mikrowellen- und Millimeterwellenradarsysteme für Automobil-, Sport- und Sicherheitsanwendungen zur Entfernungs- und Geschwindigkeitsmessung eingesetzt. Dies wird durch eine zunehmenden Reife und Verfügbarkeit von integrierten Hochfrequenzschaltungen und -Komponenten ermöglicht, was solche Systeme kompakter und gleichzeitig kostengünstiger werden lässt. Das wachsende Interesse an solchen Sensoren besteht vor allem aufgrund der Vorteile die Millimeterwellenradare gegenüber anderen entfernungsmessenden Sensoren bieten, da ihre Leistungsfähigkeit weniger von Staub, Nebel, Regen, Schnee oder dem Umgebungslicht abhängt. Bei heutigen Millimeterwellenradarsysteme ist eine der wichtigsten Anforderungen an eine hohe Integration und Miniaturisierung des Gesamtsystems. In diesem Zusammenhang spielen Millimeterwellen-Radarsysteme die oberhalb von 100 GHz arbeiten wegen den kleinen Antennenabmessungen aufgrund der kurzen Wellenlänge eine wesentliche Rolle. Dies ermöglicht eine hohe Integrationsdichte durch Kombination der integrierten Schaltung (engl.: Integrated Circuit (IC)) mit der Antenne in einem Chip-Gehäuse. Eine große Herausforderung bei der hohen Integrationdichte ist jedoch die Prüfung der integrierten Millimeterwellen- Schaltungen. Aufgrund der hohen Frequenz, fehlender oder sehr kleiner und schwierig zu erreichender Kontaktflächen (Pads), ist der Zugriff auf die Komponenten für kontaktbasiertes Testen eingeschränkt. Um das Problem zu überwinden hat sich ein sog. Selbsttest (engl.: Built-in-Self-Test (BIST)) -Ansatz als eine sehr effiziente Testmethode erwiesen.

Inhalt der Arbeit sind der Entwurf und die Realisierung mehrerer hochintegrierter Millimeterwellen- Radare für Nahbereichsanwendungen im 122 GHz Frequenzband. Zuerst werden dazu typische Systemarchitekturen von Millimeterwellen-Radarsensoren, insbesondere Homodyn- und Heterodynsysteme, sowie monostatische und bistatische Systeme vorgestellt. Weiter werden Antennen, sowie Aufbau- und Verbindungstechniken mit dem Fokus auf Nahbereichsanwendungen diskutiert.

Im ersten Teil wird ein hochintegrierter Single-Chip, Dual-Channel-Radar-Sensor-System aufgebaut und untersucht. Das Vorgehen bei Entwurf und Implementierung wird von der Systemebene ausgehend dargestellt. Anschließend werden die Designüberlegungen und die Charakterisierung der einzelnen Komponenten bis zur vollständigen Umsetzung der vorgeschlagenen Architektur kurz beschrieben. Dabei werden Zweikanal- On-Chip-Antennen bei 122 GHz eingesetzt und deren Funktionalität demonstriert. Durch diese wird eine Abstands- und Winkelinformation mit dem gezeigten Radarsensor erreicht.

Im zweiten Teil wird die Funktionsweise von heterodynen Radarsystemen, die also mit einer Zwischenfrequenz arbeiten, dargelegt. Durch diesen Ansatz können einige der bei homodynen Radarsystemen häufig auftretenden Probleme überwunden werden. Dies sind insbesondere Gleichspannungs-Offsets am Empfängerausgang durch Reflexionen in kurzer Entfernung, die sich aus der nicht-idealen Hardware ergeben. Die Messergebnisse zeigen, dass eine bessere Leistungsfähigkeit in der heterodynen Konfiguration erreicht werden kann, insbesondere im Vergleich mit ähnlichen Arbeiten, die zuvor veröffentlicht wurden. Des weiteren wird der Einfluss von Temperatur, zufälligen Störungen und Rauschen auf die Leistungsfähigkeit des Sensor untersucht. Schließlich wird auch das Problem des dynamischen DC-Offsets untersucht.

Im letzten Teil dieser Arbeit wird erstmalig eine im Sensor integrierte Autokalibrierungsmethode und ein neuartiger eingebauter Selbsttest für Radarsysteme vorgestellt und untersucht. Das Testen von modernen komplexen elektronischen Systemen ist ein sehr kostspieliges Verfahren, weshalb spezielle Strukturen zur Vereinfachung dieses Prozesses bereits während der Entwurfsphase in das System eingefügt werden. Dafür wird eine Zweikanal-Heterodyn-Radararchitektur mit integrierter Selbsttest- und Kalibrierfunktion vorgeschlagen. Eine Methodik für die eingebettete automatische Kalibrierung und das integrierte Testkonzept wird anschließend beschrieben. Um das vorgeschlagene Verfahren zu verifizieren wird ein hochintegrierte Radarschaltung bei einer Frequenz von 122 GHz realisiert und das Konzept unter Verwendung eines darauf basierenden Prototyps mittels Messungen verifiziert. Darüber hinaus wird zum ersten Mal die Messung der komplexen S-Parameter einer an die integrierte Schaltung angeschlossenen Millimeterwellenantenne demonstriert.

Contents

List of Abbreviations and Symbols

List of abbreviations

AC	Alternating Current
ADC	Analog Digital Converter
AiP	Antenna in Package
AoC	Antenna on Chip
ATE	Automatic Test Equipment
AUT	Antenna Under Test
BEOL	Back End of Line
BGA	Ball Grid Array
BPF	Band Pass Filter
BiCMOS	Bipolar-Complementary Metal-Oxide-Semiconductor
BiSC	Built-in-Self-Calibration
BIST	Built-In-Self-Test
BW	Band Width
CP	Charge Pump
DAC	Digital Analog Converter
DC	Direct Current
DfT	Design for Test
DoA	Direction of Arrival
DR	Dynamic Range
DSP	Digital Signal Processing
DUT	Device Under Test
EM	Electromagnetic
eWLB	Embedded Wafer Level Ball Grid Array
FFT	Fast Fourier Transformation
FMCW	Frequency Modulated Continuous Wave
FPGA	Field Programmable Gate Array
FSL	Free Space Loss
GCPW	Grounded Coplanar Waveguide
GSG	Ground-Signal-Ground
HBT	Heterojunction Bipolar Transistor

IC	Integrated Circuits
IF	Intermediate Frequency
IoT	Internet of Things
ISM	Industrial, Scientific and Medical
IQ	In-phase and Quadrature
LNA	Low Noise Amplifier
LO	Local Oscillator
LPF	Low Pass Filter
LRR	Long Range Radar
KGD	Known Good Die
mm-Wave	Millimeter-wave
MMIC	Monolithic Microwave Integrated Circuit
MRR	Mid Range Radar
NF	Noise Figure
OWP	On Wafer Probe
PC	Personal Computer
PCB	Printed Circuit Board
PD	Power Detector
PFD	Phase Frequency Detector
PLL	Phase Locked Loop
QFN	Qual-Flat no-Lead
RCS	Radar Cross Section
RF	Radio Frequency
RFIC	Radio Frequency Integrated Circuit
SDR	Software Defined Radio
SRD	Short Range Device
SiGe	Silicon Germanium
SiP	System in Package
SHM	Sub-Harmonic Mixer
SMD	Surface Mounted Device
SNR	Signal to Noise Ratio
SoC	System On Chip
SPI	Serial-Peripheral-Interface
SRR	Short Range Radar
TM	Top Metal
VCO	Voltage Controlled Oscillator
VGA	Voltage Gain Amplifiers
VNA	Vector Network Analyzer

USB	Universal Serial Bus
USRP	Universal Software Radio Peripheral
ZIF	Zero-IF

Frequently used symbols

c	Speed of propagation of electromagnetic waves
c_o	Speed of light
d	Distance of a target
DR	Dynamic range
f	Operating frequency
f_{beat}	Beat frequency
f_D	Doppeler frequency
f_s	Sampling rate
Δf	Band width
f_D	Doppler frequency shift
G_{TX}	Transmit antenna gain
G_{RX}	Receive antenna gain
NF	Noise Figure
P_{TX}	Transmit power
P_{min}^{in}	Minimum detectable power
R	Target range
ΔR	Range resolution
t	Continuous time
T_{sweep}	Modulation sweep time
κ	Boltzmann constant
τ_D	Round-trip delay
ΔV_D	Doppler resolution
λ_0	Free space wavelength
Γ	Reflection coefficient
υ_R	Radial velocity
ϵ_r	Relative dielectric permittivity
ϵ_{eff}	Effective dielectric permittivity
σ	RCS of a target

1 Introduction

RADAR, which is an acronym for **RA**dio **D**etection **A**nd **R**anging, has a rich history dating back to Heinrich Hertz's classical experiments in the 1880's, [Sko01]. The first radar system was invented by Christian Hülsmeyer (1881-1957) in 1904 to avoid vessel collisions on the river Rhine even in bad weather conditions. The history of millimeter-wave (mm-wave) started in the 1890s, [Wil84]. By that time Hertz had performed experiments confirming Maxwell's theory predicting the possibility of radio waves.

Today, radar systems exist for a variety of applications from weather observation to security systems, from automotive applications to consumer electronics. In the last decade, a strong growth of highly integrated ICs (Integrated Circuits) in radar sensor technologies are seen. This expansion has been fueled by the tremendous technology advances on silicon technology that achieve high integration, low cost and low power sensor circuits for large volumes, especially at mm-wave frequencies. In the future, this trend is set to continue at an accelerated pace given the emerging opportunities and applications in single-chip mm-wave radar made possible by silicon technologies.

Automotive radar is one of the current challenges driving innovation in small low cost mm-wave radar. Today, automotive radar systems are widely installed in many transportation vehicles to assist driving safety and comfort. These systems provide drivers with information about the distance between the vehicle and other vehicles, as well as obstacles present on the road. The radar systems can take a further step into automatically controlling the accelerate/brake system of the vehicles to avoid any accident.

Advances in silicon semiconductor technology and the proliferation of automotive radar sensors has brought a new drive to small and highly integrated low-cost radar sensors, which can be used for speed and distance measurement in commercial and consumer applications. Interest in consumer applications of radar is of course not new. In 1993 the IEEE held a symposium on this topic and there are already small low-cost hand held Doppler radar systems on the market, [IEE93]. Another important driving factor in mm-wave sensing is the new era in Internet of Things (IoT) remote sensing, which consists of acquiring information about an object without making physical contact with it. In these application areas, the object can be nearby or several meters away.

The main research objective of this dissertation is the development of mm-wave radar sensor system concepts for short range applications. The developed radar sensor systems use the measurement principle of radar for a wide range of applications. The primary focus lies on the system-level investigation of highly integrated single chip radar transceivers at mm-wave frequencies. The main requirements of mm-wave radar based sensor systems, namely, miniaturization, low-cost system, and reliability are briefly addressed. A new art of system level Built-In-Self-Test (BIST) and embedded radar calibration methodologies are proposed and implemented.

1.1 Radar sensors at millimeter-wave frequencies: State-of-the-art

The mm-wave frequency range offers several GHz of bandwidth that can be employed for bandwidth-intensive applications such as data communication, high resolution radar sensor applications, remote sensing and imaging sensor systems for safety. However, the mm-wave and sub-terahertz frequencies have not been yet fully explored in integrated circuits and systems. Many IC prototypes have been demonstrated at the frequencies between 30 GHz and 100 GHz for radars, imagers, and communications.

In this section, an overview of the technological advances in mm-wave circuit components and sensors is presented. Table 1.1 revises the most frequently used application areas of mm-wave band. The employment of the 57-64 GHz band for communication technology has been very important for the development of mm-wave technologies, as the data rates and bandwidth provided by this band are excellent. The 60 GHz frequency band has been utilized and standardized for data communications specially in the ultra-broadband wireless indoor applications. At the time of this writing, the cellular industry is just beginning to explore application beyond "‘4G"’ and "5G"’ by the industry. On October 22nd 2015, Federal Communication Commission (FCC) proposed new rules (FCC 15138) for wireless broadband frequencies of 28 GHz,

Applications	Frequency band
5G cellular	28 / 38 GHz
Broadband wireless data link	57 - 64 GHz
Automobile radar	76 - 81 GHz
Imaging	94 / 183 GHz
Short range devices	122 - 123 GHz
Radiometry and astronomy	183 / 220 GHz

Table 1.1: Overview of mm-wave applications

37 GHz, 39 GHz and 64-71 GHz bands. Researchers are targeting these frequencies for 5G applications and IoT (Internet of Things). In addition to communication application, recently the 60 GHz mm-wave band has been used for human mobility tracking, [ZPYM16], and hand gesture sensing sensor developed by Google and Infineon [NJB$^+$16].

The 77-81 GHz band has been licensed for automotive radar, which is one of the driving innovation in low cost millimeter-wave radar. First offered in 1999 by Mercedes-Benz in their luxury cars, a wide range of systems operating at 77 GHz is now available from a wide range of vehicle manufacturers. Based on 77 GHz MRR (Mid Range Radar) and LRR (Long Range Radar) the latest sensors can offer functions of parking aids, blind-spot monitoring, automatic cruise control, collision avoidance, pre-collision intelligence, night and fog vision, and road surface monitoring. Today, with the rapid evolution of SiGe, and CMOS technologies, [LDH$^+$13, LLHH10a], 77 GHz automotive radar transceivers can be realized in a single chip at low cost and compact size which will further facilitate the use of such systems in low and medium-price segment cars [FKJ$^+$08, NYP$^+$08, TDG$^+$09, HWS$^+$07].

Frequencies above 100 GHz are considered more useful for mm-wave imaging because their shorter wavelengths can offer higher resolution. In recent works, super resolution sensors have been investigated and demonstrated in the mm-wave band [CCYH04]. As the operating frequency increases, it becomes increasingly difficult to design and implement efficient and robust integrated circuits because of the limited performance of devices, the process variations, and inaccurate measurements and device models. The FCC frequency allocations contain an ISM (Industrial, Scientific, Medical) band at 122-123 GHz. From a practical point of view, the D-band (110-170 GHz) poses a good compromise between the available circuit performance in silicon technology and practical antenna size. In particular, the 122-123 GHz frequency band was designated to be used for short range devices in a large number of countries, worldwide.

In addition, sensors in the mm-wave spectrum offer a contact-less inspection of materials with radiation and high spatial resolution, which are used for imaging and security applications, [ASG+12]. For example a mm-wave scanner, which is a whole body imaging device used for detecting objects concealed underneath a person's clothing using an electromagnetic radiation, is becoming popular. Typical uses for this technology include detection of items for commercial loss prevention, screening at government buildings and airport security checkpoints. The allocation of new spectrum has increased in the last decade as mm-waves are now practical and affordable, and they are finding all sorts of new uses. This step takes the pressure off the lower frequencies and expands the band-width intensive applications in the future to create new consumer products at high frequencies and greater data rates. While this new spectrum gives some expansion, there are some limitations using frequencies in the mm-wave band.

One of the key limitations of mm-waves is the limited range. The free space loss is calculated with $L_{FSL} = (\frac{4\pi R}{\lambda})^2$, where R is the line-of-sight distance between the transmit and receive antennas and λ is the wavelength in meter. Mathematically, the free space loss increases with frequency. However, the equation assumes unity gain antennas at each end. A larger antenna is required to get the same gain at a lower frequency. Increasing the operation frequency means operating at shorter wavelengths, allowing either smaller antenna dimensions or a larger electrical aperture at the same mechanical dimensions.

Atmospheric attenuation effect and increased free space path loss, and decreased signal penetration through obstacles should be taken into account. The mm-wave range encompasses different absorption features, which are required to properly design a reliable mm-wave system that rely on propagation through the atmosphere. Compared to the negligible attenuation at lower frequencies, signals above 100 GHz are subject to high oxygen absorption and rain fade of RF energy. Thus, these bands cannot support long-distance communications, but are well suited for short-range applications. The set of curves in Figure 1.1 shows the expected atmospheric path loss as a function of frequency, in the case of dry air, fog and in the presence of rain fall. Consequently, in a dry air, the attenuation of mm-wave transmission for short range application is not a matter of prime concern.

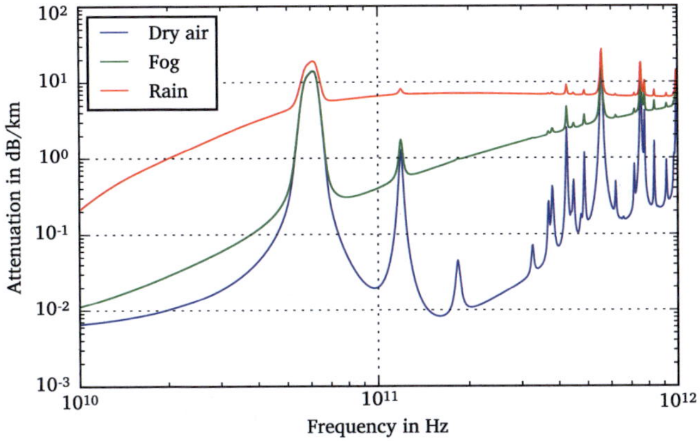

Figure 1.1: Approximate atmospheric attenuation in dry air in the mm-wave spectrum in the frequency range of 10 GHz to 1 THz, [HJL87].

1.2 Radar sensors at 122 GHz

The attractiveness of mm-wave frequencies from an application perspective is tempered by the costs associated with semiconductor device technologies and miniaturization. The short wavelength at mm-wave frequencies permits compact on-chip passive components design, facilitating the implementation of fully integrated transceivers and reducing cost. The ISM band at 122 GHz spectrum is a good compromise for short range radar based distance, angle and speed sensing, in terms of integration and miniaturization. Low-cost solutions in SiGe-BiCMOs technology is possible with mm-wave SoC approach. The possibility of an integrated antenna on-chip or in package becomes viable, thus avoiding the cumbersome and high cost packaging solutions that would be required at mm-wave frequencies. Lastly, as illustrated in Figure 1.1, the atmospheric attenuation at 122 GHz is insignificant, even in high humidity, indicating that the environmental conditions will not affect the correct operation of the system. The available bandwidth of 1 GHz allows an object distance separation of 15 cm, thus for short-range operation only single target scenarios are feasible. The maximum allowed Effective Isotropically Radiated Power (EIRP) of +20 dBm translate to a transmitter output power in the order of 0 dBm, if a high gain antenna with 20 dBi antenna gain is assumed.

Short range radar applications

A large spectrum of applications for non-contact distance, speed and angle-measurement sensors can be envisioned. Application areas include consumer goods, industrial engineering, process automation, automotive engineering etc. Consumer applications include hand-held distance measurement. In automation and industrial applications work piece monitoring and safety installations in manufacturing lines are promising examples. For home or public building infrastructure mm-wave sensors can be used for automatic doors and alarm systems.

For the above mentioned applications, often only distance information to a well defined object is required. However, in addition to the measurement metrices, price, robustness, sensor size and measurement reliability are additional requirements. Furthermore a distance sensor based on electromagnetic waves exhibits distinct advantages compared to other measurement principles like laser, ultrasonic or optical triangulation. It is robust and reliable environment, influences like ambient light, sound, vibration and temperature do not impede the measurement or can be mitigated. In addition, mm-waves are invisible to the human eye and the sensor itself can be fitted behind a cover. Another advantage of mm-wave radar sensors is the fast measurement cycle, depending on the distance range and accuracy, the measurement can be performed with nearly arbitrary speed.

1.3 Requirements of radar sensors for short range applications

Mm-wave sensors requires a system architecture that can deliver the performance, reliability, and low cost by using design techniques and available semiconductor technologies. The benefit on an increase in frequency not only manifests in the ability to use more compact high gain antennas or even on-chip integrated antennas, but additionally results in higher accuracy for distance measurements with micrometer accuracy. The main requirements of future portable sensors and devices are summarized as follows.

Miniaturization The proliferation of radar sensor electronics into portable devices for consumer electronics, automobiles, and industrial applications is forcing the cost and further miniaturization of the sensor electronics. As the costs of integrating more and more components onto the same piece of silicon increase, alternative ways to integrate the whole system into a single package, such as surface mount technologies, are becoming more attractive. In the case of mm-wave radar sensors, integration of the silicon IC and the antennas in the same housing plays a crucial role in terms of miniaturization. Increasing the frequency of operation will allow further miniaturization of key components and offer small-size solutions.

Accurate, short range measurement capability Despite several successful demonstration of compact, low cost radars in recent years, reported systems tend to show limitations towards measurement and detection of short range objects. Frequency modulated continuous wave (FMCW) radars transmit and receive simultaneously and are capable of very short minimal measurement ranges. In practice, the presence of a large DC component in the demodulated IF, which must be filtered to prevent amplifier saturation, limits the minimum measurement range in an FMCW system. Most FMCW type radars presented so far, implement a homodyne architecture, which introduces problems with DC offsets that reduce sensitivity in the near range. Thus, these systems require some additional and often advanced calibration and signal processing steps. At the outset of this PhD thesis a heterodyne radar system architecture is investigated to improve the short range measurement capability. Based on theoretical investigations on various transceiver architectures, a heterodyne architecture could overcome the DC-offset issues at short range.

Test cost reduction Testing cost is a significant part of the overall product cost for modern integrated circuits. In particular, testing RF and mixed-signal components in a System on Chip (SoC) to examine their conformance to specifications [CC10] could account for up to 70% of the overall cost of the SoC [PDAO10]. In addition to random defects and systematic failures that could result in defective devices, parametric variations in circuit, device and process parameters cause deviations in device performances, such as gain, power, frequency of operation and bandwidth. This necessitates the need for a new approach to RF and mm-wave design through integration of on-chip BIST and Built-in-Self-Calibration (BiSC) circuits. Reducing RF test complexity and cost is still an open research topic that has been addressed in a number of different approaches. This is achieved using digitally assisted mm-wave circuits to monitor and calibrate a mm-wave shortcomings in integrated ICs. A BIST and embedded calibration concept for an integrated D-band radar sensor is addressed in this thesis.

1.4 Goal and organization of this thesis

This dissertation investigates concepts of miniaturized radar sensor systems at 122 GHz operating frequency with a particular emphasis on short range sensing applications. The focus is to build-up a low-cost, highly integrated, single chip radar system for universal application. System design considerations, comparison of different radar system architectures and integration concepts for miniaturization are analyzed briefly. The broad theme of this thesis is to explore the most appropriate and feasible solution in terms of system architecture of sensors above 122 GHz for short range distance and angle sensing applications. Furthermore, a Built-In-Self-Test concept for a highly integrated radar sensor is realized and analyzed.

The tasks and organization of the thesis can be broken up as in Figure 1.2. In Chapter 2, the working principles and definitions of radar system metrices are discussed. Types of transceiver architectures, choices of antenna and packages, with a focus on mm-wave technologies is addressed in details. Basic concepts pertaining to FMCW radar sensors are reviewed. Chapter 3 is devoted to a single chip miniaturized dual channel radar sensor with on chip antennas to measure distance and estimate direction of a single target. The system design and architecture of the radar sensor are described. The miniaturization concept and integration of the on-chip antennas are discussed. Measurements of the transceiver system and performances are presented in this chapter.

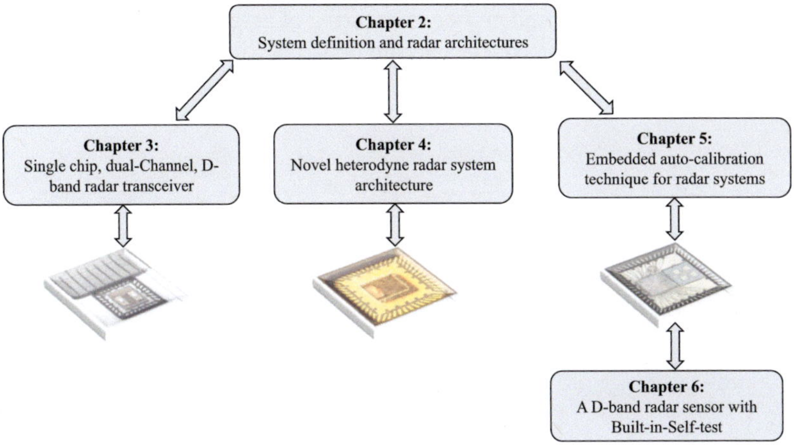

Figure 1.2: General outline of this thesis.

Chapter 4 describes and discusses the design approach of a fully integrated heterodyne radar system architecture operating at 122 GHz. This proposed radar transceiver achieves a very good performance in terms of DC-offset eliminations for short range sensing applications. The radar circuit description, implementation of the system and measurement results are discussed. Furthermore, this chapter compares the performance of homodyne and heterodyne radar architectures for short range distance sensing.

An embedded auto-calibration technique for radar systems is introduced in Chapter 5. A brief summary of the status and trends of mm-wave testing methodologies are revised in Chapter 6. The concept is based on a monostatic heterodyne radar system architecture which is built symmetrically, with one of the channel connected with antenna and the second channel terminated by a programmable impedance tuner for built-in calibration scheme. The concept is verified by comparing with on-wafer probe measurements. A systematic error calibration concept for RF-impairments in radar systems is investigated. The same sensor architecture with several Built-In-Self-Test capability is presented, which shows a low-cost, low frequency test methodology at mm-wave frequencies. The integrated components could be accessed and monitored digitally during BIST mode. Furthermore, a novel method for measurement of a complex S-parameter and on-line monitoring of antenna matching is shown for the first time. Finally, conclusions are drawn in Chapter 7.

2 System definition and radar architectures

This chapter reviews the theory and working principles of radar sensor systems. It starts with a basic concept and definition of radar performance metrics. The system level considerations and operation of radar relevant to this thesis are explained. Section 2.3 discusses different radar system architectures that can be realized in a standard integrated circuit form and the main advantage and disadvantages of each architecture is analyzed. The working principle of FMCW radar is described in section 2.2. Mm-wave antenna concepts, packaging and interconnect technologies, in the context of this work are revised in the final section.

2.1 Radar performance metrics

A radar system consists of a transmitter and a receiver, an antenna, a controller and of digital processing software that interprets the results to obtain distance, velocity and angle information. There are several parameters that quantify the performance of a radar system. The degree to which each parameter needs to be satisfied depends largely on the application. For example, the case of a single large nearby target has vastly different requirements from that of an airborne radar that has to detect extremely fast moving objects at distances of several hundred kilometers.

Radar range equation

The basic principle of radar is that an RF signal is transmitted towards the target of interest, which is reflected from the target and then received by the radar antenna. Information regarding the distance, relative speed and angular position of the target is detected using the reflected signal. The received power can be described by the radar equation [Sko01] as:

$$P_{Rx} = \frac{P_{TX} G_{TX} G_{RX} \lambda^2 \sigma}{(4\pi)^3 R^4} \tag{2.1}$$

where P_{Rx} is the received power, P_{Tx} is the transmitted power, G_{TX} is the transmit antenna gain, G_{RX} is the receive antenna gain, λ is the wavelength of the carrier frequency, σ is the RCS of the target, and R is the range of the target, [Sko01]. Note that this radar equation is valid only if the target is in the far-field of the antenna.

Velocity estimation

The Doppler effect was discovered by Christian Doppler in 1842. For a moving target, its velocity is measured based on the well-known Doppler effect. If the radar transmits a signal at a frequency f_o, the reflected signal from the moving target is subjected to a Doppler frequency shift f_D from its transmitted frequency f_o induced by the relative motion between the radar and the target. In the case where a target has a radial velocity v_R, the Doppler frequency shift f_D is determined by the radial velocity of the target and the radar transmitted frequency f_o:

$$f_D = -2 f_o \frac{v_R}{c} \tag{2.2}$$

where v_R is defined as positive value when the target is moving away from the radar.

Maximum range

The maximum range of a radar, R_{max}, is the maximum distance where a target can be detected, which depends primarily on the SNR of the receiver. Assuming the target to be in the center of the antenna beam, the maximum radar detection range derived from equation2.1 is defined as:

$$R_{max} = (\frac{P_{TX} G_{TX} G_{RX} \lambda^2 \sigma}{(4\pi)^3 P_{min}^{in}})^{\frac{1}{4}} \tag{2.3}$$

where P_{min}^{in} is the minimum detectable power at the input of the receiver. The expression demonstrates the relationship between the target range, transmitted power, and minimum detectable received power. Increasing the transmitted power and/or decreasing the minimum detectable received power increases the maximum range of the radar.

Receiver sensitivity and dynamic range

The minimum receiver power, which can be detected by the radar receiver is refered as receiver sensitivity. To estimate the sensitivity of a radars systems, the parameters in the radar equation 2.1 are used, [Sko01]. Independent of any signal modulation schemes used, the sensitivity can be described as:

$$\frac{S}{N}_{min} = \frac{P_{RX}}{P^{in}_{min}} = \frac{P_{\bar{T}X} G_{Tx} G_{RX} \lambda^2 \sigma}{(4\pi)^3 k T_{Sys} \Delta f\, NF\, R^4} \tag{2.4}$$

$$P^{in}_{min} = k\, T_{sys}\, \Delta f\, NF \tag{2.5}$$

where $P_{\bar{T}X}$ is average transmit power, k =$1.38 \cdot 10^{-23} Ws/K$ is Boltzmann constant, T_{Sys} is temperature of the system, BW is receiver bandwidth, and NF is noise figure. Therefore, the minimum detectable signal or sensitivity of a radar sensor, apart from the antenna gain and the noise figure which couldn't be influenced by signal modulation scheme, depends mainly on the average transmit power.

Minimum detectable signal depends on the receiver bandwidth (Δf), noise figure (NF), temperature (T_{Sys}) and required signal-to-noise ratio (SNR). A narrow bandwidth receiver will be more sensitive than a wider bandwidth receiver.

$$P_{min} = k T_{Sys} \Delta f\, NF (S/N)_{min} \tag{2.6}$$

Another standard receiver specification is the dynamic range, which is the ratio of maximum-capable to minimum detectable signal amplitude. The ratio of the largest receivable signal to the minimal detectable signal is defined as follows:

$$DR = \frac{P^{in}_{max}}{P^{in}_{min}} \tag{2.7}$$

where DR is the dynamic range and P^{in}_{max} is the maximum allowable power at the input of the receiver. The minimum detectable signal, P^{in}_{min}, must be above the receiver noise level. The DR of the system will affect the maximum range at which a target can be detected.

In order to achieve a high dynamic range in FMCW modulated radar sensors two effects should be considered. Primarily, any noise in the receiver will lead to a higher noise level in baseband, which depends on the amplification level the receiver. For example, in the presence of a DC-offset, which will be detailed in Section 4.1, a close-by object can be masked by noise and will fail to be detected by the radar. The second effect is false detection of an object when the receiver is saturated due to high reflection and signal harmonics with frequency components near to baseband. To improve the dynamic range, a high-pass filter can be implemented to compensate the baseband signal. However, a full compensation is not possible, due to the coupling of range and speed of the range superimpose in baseband for FMCW.

Range resolution and accuracy

The range resolution defines the ability of resolving two point-targets within the same antenna beam, close together in the range domain. Because the delay-time τ of a radar signal returned from a target is related to range R by $\tau = \frac{2R}{c}$, the resolution in range is directly related to the resolution in delay time. The range resolution of FMCW radar is determined solely by the bandwidth of the chirp, independent of the chirp duration which permits fine range resolution to be achieved. The range resolution ΔR is described by:

$$\Delta R = \frac{c}{2\Delta f} \tag{2.8}$$

where c is the speed of light and Δf is the signal's bandwidth. The range resolution is independent of the center frequency of operation, although there is a clear advantage in operating at higher frequencies due to the higher bandwidths available.

The accuracy of the radar is the average absolute error in the detected distance R. It is represented by the standard deviation, δR, and depends on the SNR, as well as, on systematic errors such as the non-idealities of the radar waveform. Obviously, due to the increasing free space loss with distance, the accuracy of a radar degrades with increasing distance to the target.

$$\delta\, R \approx \frac{\Delta\, R}{\sqrt{2\, SNR}} \approx \frac{c}{2\Delta\, f\sqrt{(2\, SNR)}} \tag{2.9}$$

where $\Delta\ R$ is the range resolution. A high accuracy FMCW radar sensor at mm-wave frequency is reported in [Sch16].

Doppler resolution

Doppler resolution refers to the ability of resolving two targets in the radial velocity. The Doppler velocity ΔV_D is related to the measurement time T and the carrier frequency by:

$$\Delta V_D = \frac{c_o}{2\, f_c\, T} \tag{2.10}$$

Angle measurement

In addition to distance and relative speed measurements, determining the direction to a target is one of the basic functions of a radar. The most widely used methods of determining the direction to a source of signals are methods of linear scanning and sequential switching of radiation patterns, which are performed by single channel direction finders. Figure 2.1.a illustrates the schematic of the principle. Based on these methods, the direction to a source was determined by a comparison of signals received sequentially by antennas with various radiation patterns.

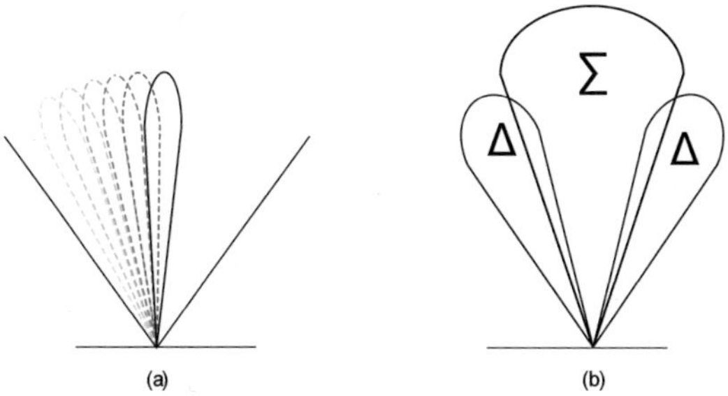

Figure 2.1: Angle measurement methodology with radar sensors: (a) Scanner principle for angle determination, with a closely concentrated beam scanning the whole area to detect a target. (b) Monopulse principle.

Another common principle to find a direction of an object is monopulse method. The direction finding lies in the reception of signals reflected from a target simultaneously along several independent receiving antennas with the subsequent comparison of their parameters, as depicted in Figure 2.1 .b. The receiving antennas may differ due to the beam characteristics or of the horizontal position displacement. There are two different kinds of monopulse radars:

- **Amplitude-comparison monopulse** consists of multiple antennas building the individual signals and giving them a common phase center. By comparing the relative amplitude of the pulse in the two beams, its position in the beams can be determined.
- **phase-comparison monopulse** uses multiple separate antennas looking all in the same direction with a separation baseline resulting in independent phase centers.

For two or more antennas separated by a distance d, with a wavefront incident angle θ, then the extra path, Δl, the signal must travel between antenna 1 and antenna 2 results in a phase difference, $\Delta\Phi$, between the two antennas.

$$\Delta l = dsin(\theta) \tag{2.11}$$

Figure 2.2 shows a schematic representation of the angle measurement principle with n parallel receiving channels. The path difference results in a phase

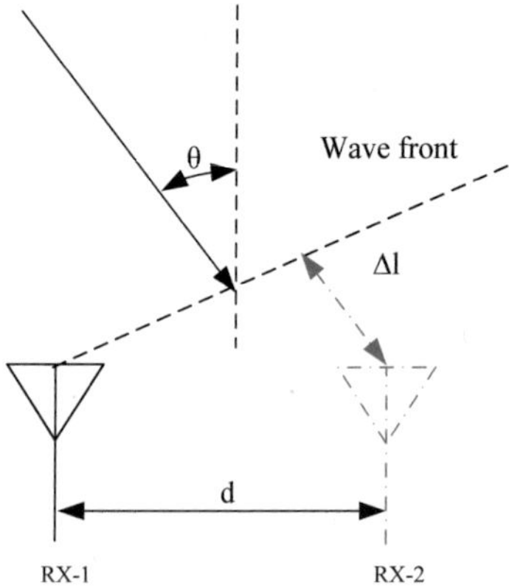

Figure 2.2: Principle of angular measurement based on phase difference in the azimuth plane by two receiving antennas.

difference $\Delta\Phi$ between the signals from the two antenna:

$$\Delta\Phi = 2\pi\frac{\Delta l}{\lambda} = 2\pi\frac{dsin(\theta)}{\lambda} \tag{2.12}$$

This can be used to calculate the direction of arrival using:

$$\theta = asin\left(\frac{\lambda\Delta\Phi}{2\pi d}\right) \tag{2.13}$$

Due to the azimuth position of the target, a linear phase shift can be observed in the receive signal between the consecutive antennas. The phase shift can be exploited in order to obtain the corresponding angle and hence the DoA. For a distance d = $\frac{\lambda}{2}$ between the two antenna elements, and according to the spatial sampling theorem, azimuth angles in the range of $\theta = \left[-\frac{\pi}{2}, \frac{\pi}{2}\right]$ can be measured. For unambiguous results, the antennas should be spaced half a wavelength apart, or less. This will manifest as additional lobes with the same height as the main lobe. These lobes are called grating lobes and they introduce ambiguities in terms of the position of the maximum peak in the beam pattern.

2.2 Frequency modulated continuous wave radar

Frequency modulated continuous-wave (FMCW) radar systems employ frequency modulation at the signal source to enable propagation delay measurements to determine the distance or speed to a target. Figure 2.3 shows the variation in frequency versus time for the transmitted and received signal, based on a simplified linear chirp. The frequency modulation causes a change on the transmit frequency over time, which results in a frequency difference between transmit and receive signals. The mixing process is carried out by the mixer which produces summation and subtraction frequencies of two signals. The frequency difference between two signals is called as a beat signal which is proportional to the distance between the radar and the target as well as doppler of the target.

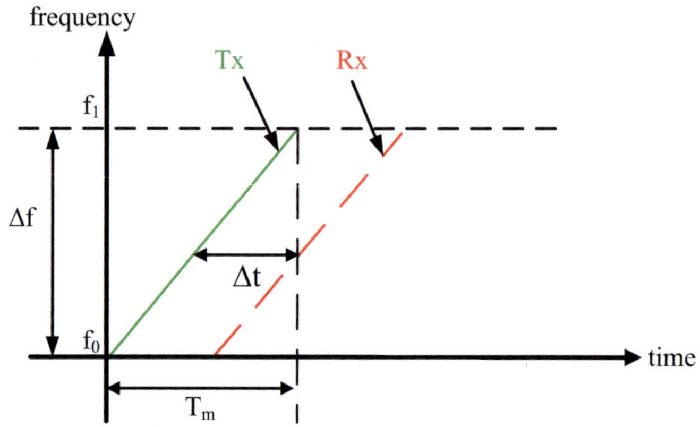

Figure 2.3: Time dependent behavior of transmitted and received FMCW signal: solid curve represents transmitted signal; dashed curve represents the received echo signal.

In FMCW radar signal processing, when there is no doppler shift in the transmitted signal or stationary target, the beat frequency measures the target range. Thus, f_{beat} can be computed as follows.

$$S_{TX}(t) = \hat{A}_{TX} \cdot cos(2\pi(f_{TX}(t)) \cdot t) \tag{2.14}$$

$$f_{TX}(t) = f_o + \frac{\Delta f}{T_m} \cdot t \tag{2.15}$$

The transmitted signal S_{TX} will be reflected from a target at a distance R and will be received by the receive antenna. The time delay of the signal will be calculated as

$$\Delta t = \frac{2R}{c_o} \tag{2.16}$$

The received signal and its frequency is thus the combination of equation 2.14 and 2.15, which is described as

$$S_{RX}(t) = \hat{A}_{RX} \cdot cos(2\pi(f_{RX}(t) \cdot t) \tag{2.17}$$

$$f_{RX}(t) = f_{TX}(t)(t - \Delta t) = f_o + \frac{\Delta f}{T_m} \cdot (t - \Delta t) \tag{2.18}$$

Since the frequency changes linearly over time, applies

$$f_{RX}(t) = f_o + \frac{\Delta f}{T_m} \cdot (t - \Delta t) = \underbrace{f_o + \frac{\Delta f}{T_m} \cdot t}_{f_{TX}(t)} - \underbrace{\frac{\Delta f}{T_m} \cdot (\Delta t)}_{f_{beat}} \quad (2.19)$$

$$f_{beat} = \frac{\Delta f}{T_m} \cdot \Delta t \quad (2.20)$$

When there is no doppler shift in the signal, the beat frequency in Equation 2.20 contains information which is directly proportional to the distance of an object as describing in 2.16. Thus, by combining the two equation, the distance to an object can be computed as:

$$R = \frac{1}{2} \frac{c_o T_m}{\Delta f} f_{beat} \quad (2.21)$$

2.3 Radar transceiver architectures

In this section, a basic radar system architecture is first described, and then its various configurations are briefly highlighted. The different aspects of these configurations are explained, and contrasted with their advantages and disadvantages while highlighting the more suitable setup for integration in a single chip system.

Monostatic radar architecture

Mono- and bistatic radar sensor architectures differ in the design of the transmit and receive antenna. In a monostatic radar architecture a single antenna performs both the transmission and reception of the radar signal, as shown in Figure 2.4. A coupler or a circulator isolates the transmitted and received signals. An ideal circulator theoretically provides infinite isolation between the transmit and receive path. However, circulators are impractical for integration at mm-wave frequencies, thus they are usually replaced by couplers. Nevertheless, in practice the isolation of couplers over a certain bandwidth is limited, which leads to leakage of the transmit signal to the receive path. This can cause a performance degradation of the following receiver components, such as the mixer. Furthermore, the mixing of two signals with the same frequency components results in a DC at the base band, which is knows as a DC-offset.

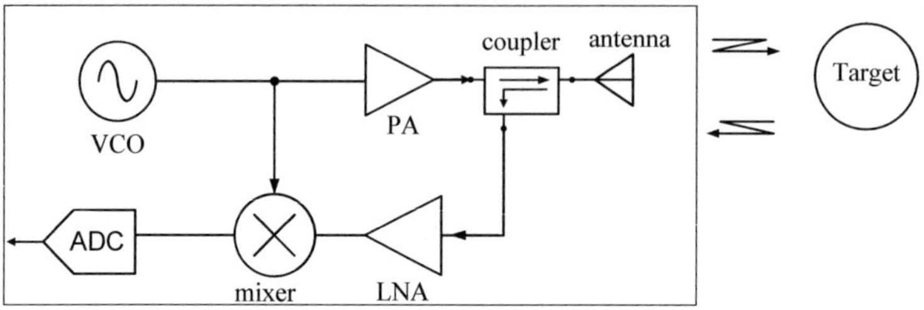

Figure 2.4: Monostatic transceiver architecture consisting of one antenna for transmit and receive a signal and a coupler for signal splitting.

The DC-offset part affects mainly the dynamic range of the Analog-Digital-Convertor (ADC) and in the worst case leads to non-detection of objects in the sight of the radar. The DC-offset topic in radar sensors is detailed in section 4.1.

Bistatic radar architecture

Bistatic radar possess spatially separated transmit and receive antenna, as shown in Figure 2.5. The IEEE standard 686 - 1997 [66198] has defined bistatic radar as a radar using antennas for transmission and reception at sufficiently different locations that the angles or ranges from those locations to the target are significantly different. There is no strict definition specifying the

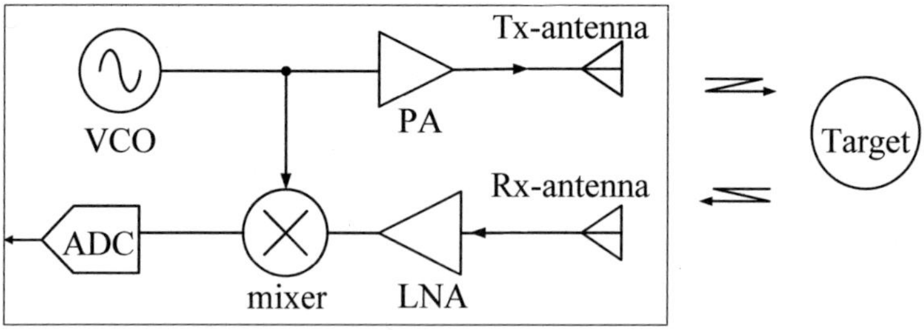

Figure 2.5: Bistatic transceiver architecture consisting of spatially separated transmit and receive antennas.

distance between the transmitter and receiver. However, several discussion suggest that the geometrical separation between the transmit and receive antennas to be larger than the distance between the respective antennas and the target, [Sie58]. Radar systems with smaller distance between the transmit and receive antennas are described as quasi-bistatic, [Jac86]. The main advantage of a bistatic sensor is the isolation between the transmit and receive path, due to the geometrical separation of the two antennas. Unlike the monostatic architecture, bistatic radars have higher costs and need additional antenna aperture area. In this work, only monostatic radar architectures are investigated.

2.3.1 Radar receiver concepts

This section examines two basic receiver concepts for radar sensor applications. The radar receiver amplifies, filters and down-converts the received target echo in such a way that the resulting intermediate frequency (IF) or a baseband signal can be processed to determine the distance and relative speed of the target.

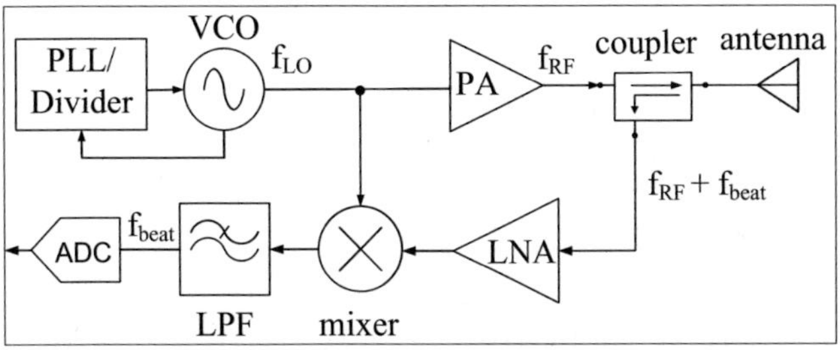

(a) Architecture of a homodyne receiver

Figure 2.6: Simplified block diagram and spectra of a homodyne radar transceiver.

Homodyne

A homodyne receiver architecture was first introduced by by F.M. Colebrook in 1924, [Tuc54]. It involves mixing the received signal with a signal at the same frequency as its carrier, so that the RF frequency is converted directly to base-band. This type of receiver, also called direct conversion, has the lowest complexity. The same signal source is used to generate the transmit signal and to supply the local oscillator (LO) signal for down- conversion to the receiver. A typical homodyne transceiver front-end architecture is shown in Figure 2.6. The main advantage of a homodyne receiver is, its simple, cheap and it does not suffer the image problem as the incoming RF signal is down-converted directly to baseband without any IF stage. The cost and power consumption of a homodyne transceiver are reduced due to the smaller number of building blocks compared to the heterodyne architecture. Performance is typically limited in a homodyne architectures in several ways. DC-offsets and $1/f$ noise are absolutely critical impairments of homodyne receivers. The major part of the DC-offset is generated at the output of the mixer when leakage from the local oscillator is mixed with the LO signal itself. This could saturate the following stages and affect the signal detection process.

Heterodyne

The principle of a heterodyne receiver architecture was first introduced by R.A. Fessenden in 1901, [Bel02]. However, it was the U.S american scientist E. H. Armstrong, who developed a superhetrodyne receiver and implement that in hardware in year 1918, [Arm24]. A heterodyne receiver down-converts the RF signal to one or more intermediate frequencies before converting it to base-band signals. A simplified block diagram of a heterodyne radar receiver is shown in Figure 2.7. The heterodyne system is the most robust transceiver architecture and has been widely used in transceivers. The benefit is the down-conversion process does not fold the RF signal back around DC, and thus do not mix frequencies located above and below the RF carrier. Any DC- error related to imperfections in the receiver electronics and $1/f$ noise, could be eliminated by using a bandpass filters (BPF) or a simple DC-block. Chapter 4 examines a D-Band single chip radar prototype with a heterodyne architecture.

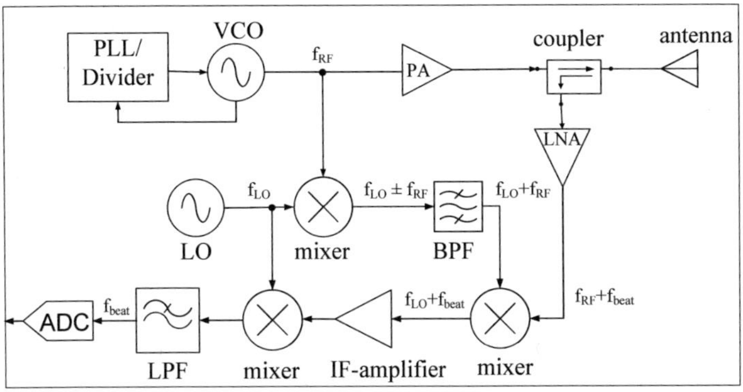

Figure 2.7: Simplified block diagram of a heterodyne radar transceiver, [Cha00].

2.4 Millimeter wave packaging and interconnects

Radar sensors that meet both cost and performance expectations require advances in how the MMICs are packaged and assembled, which must be both reliable and cost effective. This is a key element in the ability to produce truly low-cost mm-wave systems. Packages at high frequency introduce generally very important parasitic elements causing performance degradation or detuning of the MMICs characteristics enclosed in the package. The main requirements of mm-wave packages and interconnects are:

- the interconnects should provide good mm-wave performance with low reflection and low insertion loss. This is important for efficient transmission of mm-wave signals to and from the MMIC.
- passive mm-wave components, for example antennas, have to be downscaled in respect to the signal wavelength which may cause difficulties with mechanical fabrication techniques, and chip assembly processes at low cost.
- transition and feed lines to the antenna should be realized in a low-loss fashion. Therefore, the losses in the dielectric material and ohmic losses in the conductor should be reduced.

A study and investigations on mm-wave packaging and interconnects is done and analyzed in [Bee13].

Wire-Bonding

Wire bonding is by far the most used interconnect technology to connect a semiconductor chip to a printed circuit board or a package. Figure 2.8 illustrates two typical mounting scenarios: the active die places on the top of the substrate carrier, and the chip mounted in a cavity of the RF-substrate. In mm-wave systems primarily ball-wedge bonding techniques with 25-33 μm diameter gold wires and 100-300 μm wire lengths are used. The electrical performance of wirebonds depends on multiple factors related mainly to the chip mounting scheme. The wire-bonding is a mature, thermal expansion insensitive, and high yield process that supports, a well established low-cost infrastructure with high reliability. Unfortunately, the electrically long loop of the bond-wires will introduce considerable parasitic effects (about 1 nH/mm of inductance as a rule of thumb) so that they exhibit low-pass frequency characteristics in the mm-Wave range, [SM96].

(a) Chip on board

(b) Chip mounted in a cavity of RF-substrate

Figure 2.8: Chip integration with PCB antenna using bond-wire solution.

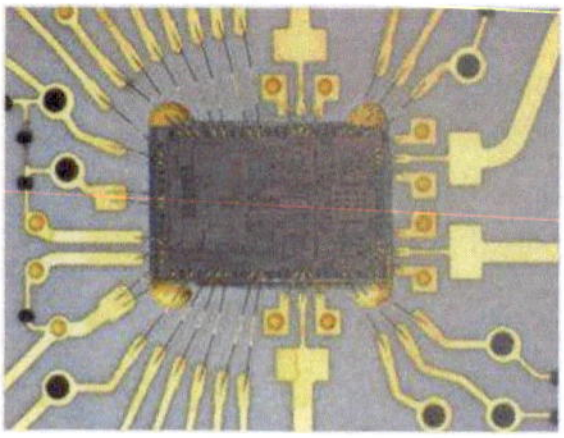

Figure 2.9: Radar MMIC bonded in a cavity of RF substrate, [HWGH10].

Flip-Chip

Flip chip is a standard processes to connect chip to a substrate. Flip Chip processes were originally developed for multi-chip applications on ceramic modules. At mm-wave frequencies, flip-chip technology has become the standard die interconnect solution for organic substrates. Flip-chip mounting has, compared to wire bond technologies, lower parasitic inductance and lower ohmic losses because of the shorter interconnect distance. Another big advantage of soldered flip-chip mounting in general is the self-aligning with high repeatability. The package carrier, either substrate or lead-frame, then provides the

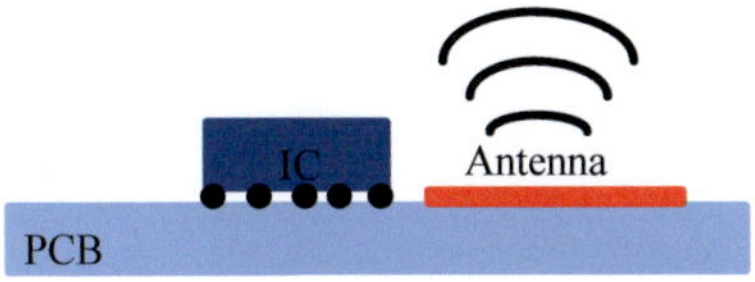

Figure 2.10: Schematic of a flip-chip mounted IC together with antenna.

Figure 2.11: Chip attach in flip-chip technology.

connection from the die to the exterior of the package. The schematic of a typical flip-chip mounting of chip is depicted in Figure 2.10. A photo of mm-wave die attached on PCB is shown in Figure 2.11. The flip-chip connection is generally formed one of two ways: using solder for chips with solder-ball or using conductive adhesive in case of stud-bumps. A bump is typically 50-60 μm high and 80-100 μm diameter. The solder bumped die is attached to a substrate by a solder reflow process. After the die is soldered, under-fill is added between the die and the substrate.

2.5 Millimeter wave antennas

Antennas are key elements and research topics in a low cost realization of mm-wave systems. For antennas at mm-wave frequencies, technical challenges lie in high efficiency, ease of integration, and high precision manufacturing. Thus, instead of a discrete antenna, namely, an antenna which is designed and fabricated independently, antenna designs tend to shift to antenna-in-package (AiP) and antenna-on-chip (AoC) solutions. The main requirements for the antenna, beyond low-cost and small size, are wide bandwidth, radiation efficiency, and technological reliability. Advanced fabrication technologies associated with low loss materials provide feasible solutions for implementing highly efficient antennas. Challenging integration issues can be solved by choosing appropriate designing methodology, manufacturing techniques and cautiously optimizing all transitions. The following section compares on-chip antennas and Antenna-in-Package for mm-wave application.

On-Chip Antenna

On-chip antennas integrated on silicon technologies have been investigated for many years in order to provide highly integrated and low-cost solution for short-range applications in the millimeter-wave bands. The main benefit of the on-chip integration is to avoid any interconnection of high frequency signals, which necessitate complex and expensive technologies. However, the drawbacks come from the poor antenna performance obtained from the low-resistive silicon. The antenna, which converts RF power from the circuits to electromagnetic (EM) radiation, finds low-resistive path through the substrate and thus incurs gain degradation. The second drawback of implementing antenna in silicon based technologies is its high-dielectric constant ($\epsilon_r \approx 11.7$), causing most of the power to be confined in the substrate instead of being radiated into free space, further degrading the radiation efficiency. To counteract this behavior, a number of techniques have been deployed to improve the radiation efficiency of on-chip antennas by redirecting the power coupled into the substrate, as in [SHBV12].

One possible method to improve the off-chip radiation of an antenna is to add an additional dielectric on top of the IC's back-end. Possible options for this are a simple superstrate, a resonator, and a lense. A 90 GHz design using a superstrate, for example, achieved a radiation efficiency of 44 - 62 %, see

[OR11]. At nearly a third of that frequency, a 35 GHz slot antenna design with metal shield in the lowest metal layer of the back-end and dielectric resonator on top of the slot achieved an efficiency of 59 %, see [NAFBSN10]. On-chip antennas including lenses were, for example, presented in [[BGK+06], [NHNA+12]]. In both cases the lense was attached at the bottom of the chip. Hence, the electromagnetic wave first has to penetrate the low resistive BiCMOS substrates. Therefore, the 77 GHz antenna design from [BGK+06] only achieved a maximum gain of 2 dBi, which is rather low for a lense antenna. The design presented in [NHNA+12], however, achieved 17.8 dBi at 180 GHz with a stated radiation efficiency of 50 %. In this work, radar sensors operating at 122 GHz with on-chip antennas are presented in Chapter 3 and 4.

3 Single chip, dual-channel, D-band radar transceiver

Silicon based technologies have been advanced towards the speed of tera-hertz, which makes it possible to build a millimeter-wave transceiver into a single chip. The high integration level of SiGe-BiCMOS technologies promises a low overall system-cost for a wide variety of new applications. In the frame work of the European Union funded project SUCCESS, [SUC13] the research teams achieved a milestone towards low-cost, miniature radar sensors all operating at 122 GHz, [SS11b, GBH+13, GPG+14]. SUCCESS targeted to develop a technology platform and best-practice design methods to enable the breakthrough of silicon mm-Wave SoCs for high-volume applications.

In this chapter, a dual-channel, short range, monostatic radar sensor operating at 122 GHz is presented for the first time. The mm-wave frontend, including two on-chip antennas is based on an integrated circuit realized in SiGe-BiCMOS technology. The concept of using a parasitic patch on an additional glass substrate above the SiGe chip was first presented in [HWGH10], which is applied for automotive radar application operating at a frequency band of 77 GHz. The realized sensor in this work is specifically tailored for short-range distance, angle and speed sensing.

3.1 Transceiver architecture and circuit implementations

Figure 3.1 shows the system architecture of the realized transceiver circuit. The realized radar MMIC is a homodyne, dual-channel sensor with two on-chip integrated antennas. To reduce the overall system complexity, a monostatic radar architecture is selected, which significantly reduces the overall size. The circuit blocks of the transceiver could be controlled digitally via SPI interface. The system level parameters of the transceiver is summarized in Table 3.1.

The monolithically integrated radar includes a VCO, power amplifiers, low-noise amplifiers, and sub-harmonic mixers (SHMs). On the transmit side, a 60 GHz LO signal is generated by a push-push oscillator to further reduce the VCO fundamental to around 30 GHz. Each of the two identical transmit channels include, a frequency doubler, a coupler and bi-directional power detector (PD) to measure the absolute output-power and antenna matching. This configuration allows to verify the functionality of the transmitter path up to the antenna by using DC measurements. An integrated directional coupler isolates the transmitted signal and the received signal reflected by the target. The coupler is designed based on an edge-coupled line structure and has a quarter wavelength at 122 GHz. The tuning characteristics and phase noise measurements of the VCO is shown in section 3.1.

A divide-by-16 chain, at approximately 3.8 GHz, is provided to an off-chip phase-locked loop (PLL), in order to lock the VCO to a stable reference frequency and modulate the transmit signal in FMCW mode. Two antennas are integrated to transmit and receive a signal simultaneously.

Type	Specification
VCO tuning range	117.5 - 123.3 GHz
Maximum transmit power	0 dBm
Operation bandwidth	$\geq$ 1 GHz
Detection range (m)	0.15 - 5 m
Range resolution	15 cm
Range accuracy	$\pm$2 mm
No. of channels	2
Antenna gain (simulated)	6.6 dBi
DC power consumption	450 mW

Table 3.1: Summary of realized radar sensor system level parameters.

In the receiver chain, the Low-Noise-Amplifiers (LNAs) are followed by two sub-harmonically pumped mixers (SHM), to mix the receive and LO signal to baseband. The two passive SHMs are driven by the same 60 GHz LO signal as the input of the frequency doubler. The baseband signals are further amplified by a pair of variable-gain amplifiers (VGA). The digital control block consists of SPI core with register files are used to control the on-chip circuits, such as

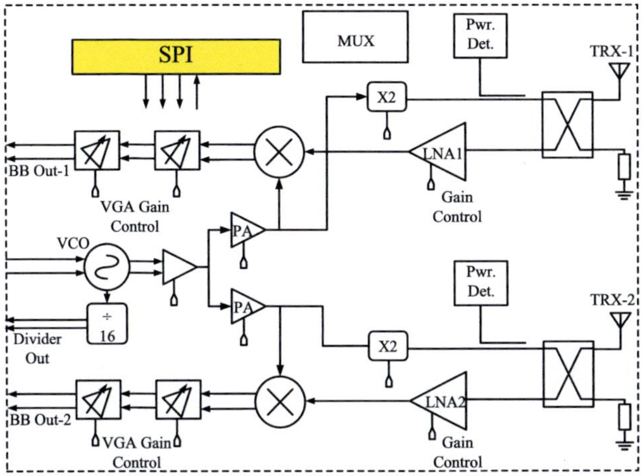

Figure 3.1: Block diagram of the realized transceiver chip

VCO coarse tuning, multiplexer output control, gain control of the PA, LNA as well as VGAs. The dual channel chip occupies a chip area of 2.5 mm x 2.6 mm, including the integrated on-chip antennas. The antennas are placed half-wavelength apart from each other. The power of the two transmitter channels is in-phase combined in free space. The circuit blocks of the transceiver have been detailed in [SS11a, SS12, SFCS12].

Semiconductor technology

A 130 nm SiGe-BiCMOS technology is adopted in this design [ADC$^+$09], which combines standard CMOS transistors and highspeed HBTs (heterojunction bipolar transistor) with relatively high breakdown voltages. Therefore, it allows to integrate high-performance SiGe front-ends and digital building-blocks into a single die to reach a very high integration level. The used technology features HBTs with 270/330 GHz f_T/f_{max} and a 5/V BV_{cbo} (HBT collector to base breakdown voltage with emitter open). There are 7 aluminum layers for interconnection, where the top-two layers have 3 and 2-μm thicknesses for constructing high-performance inductors and transmission lines. Figure 3.2 shows a simplified cross-section of the metal structure, which is used in electromagnetic modeling. A typical 50 microstrip transmission line can be constructed by using a 16-μm trace at top-metal-two (TM2) with metal one (M1) as ground. It has a loss of about 1dB per millimeter at 120 GHz.

3.2 On-chip antenna design and packaging

One of the main benefits of designing transceivers operating above 100 GHz is the small antenna size. It thus becomes feasible to integrate the antenna on the same chip with the transceiver, thereby avoiding lossy mm-wave transitions between the chip and the package. Considering resonant antennas, their physical size is proportional to the wavelength, and thus inversely proportional to the operating frequency. At 122 GHz, a half-wave length patch in free space is 1.23 mm long. An on-chip antenna is shorter because the dielectric constant of the material surrounding the antenna reduces the wave velocity by a factor of $\sqrt{\epsilon}$, where ϵ is the relative dielectric constant of the material. The on-chip

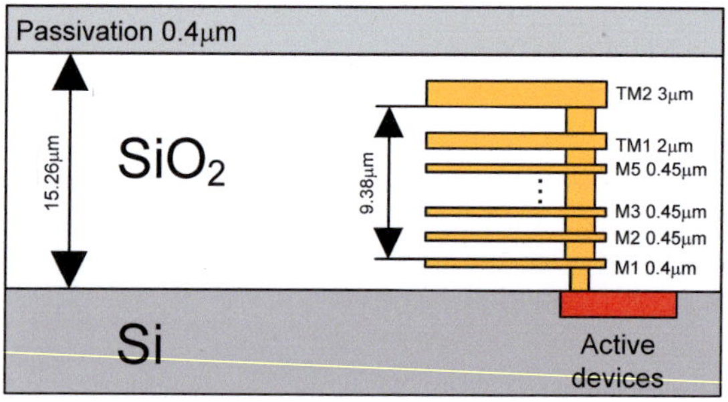

Figure 3.2: Cross sectional view of 130 nm SiGe-BiCMOS technology.

antenna of the dual-channel transceiver is designed and realized on IHP's process described in section 3.1. The on-chip antennas are implemented in the Back-End-of-Line (BEOL) process, whose schematic view is shown in Figure 3.2. The BEOL process offers 7 aluminium layers in 130 nm technology buried in the insulator of silicon dioxide (SiO_2). The top thick metal layers of TM2 and TM1 are usually used to construct RF passive components to reduce the metallic loss. All metal layers can be interconnected by stacked via arrays. The silicon substrate has a low-resistivity of 50 Ω.cm. The original thickness of the wafer is 750 μm, but it can be back-grinded to a required thickness. The

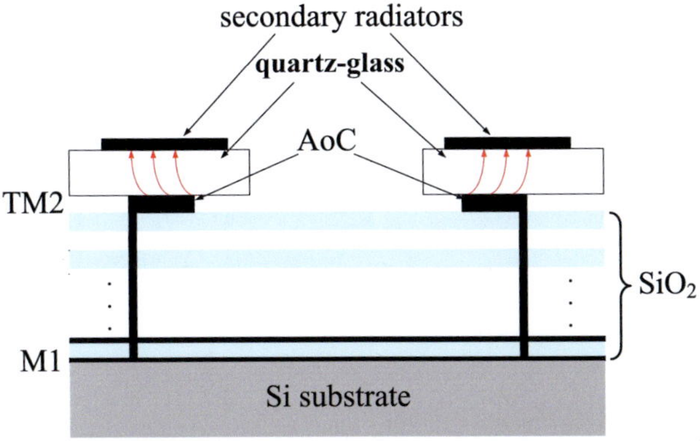

Figure 3.3: An electromagnetically coupled quartz-glass antenna placed on the top of the silicon chip. The AoC (Antenna-On-Chip) is used as a primary radiator.

proposed antenna configuration consists of a parasitic resonator placed on top of the MMIC excited by a shorted patch antenna implemented in the back-end layer stack. Figure 3.3 presents the layout of the 122 GHz electromagnetically coupled on-chip microstrip antennas. The 3- bottom metal layers (M1 -M3) are shunted together and they acts as the ground plane for the antenna, and it isolates the antenna from the lossy silicon substrate. The quarter wavelength patch is located at the top-metal layer (TM2). The $\lambda/4$-resonator is w = 160 μm and l = 290 μm in size and placed on TM2. The optimized dimensions of the antenna layout is listed in Table 3.2. The layers M1 - M3 are used as ground layer. To connect M1 to TM2 a via fence has been added.

The simulated directivity for two antenna elements in cumulative is D $\approx$ 8.76 dBi and the simulated gain is G = 6.6 dBi $\pm$0.7 dB at 120 - 123 GHz (G=ϵD, efficiency ϵ= 60$\pm$8%). Figure 3.5(b) shows the combined 3D- far field diagram of the two antennas. The simulated efficiency of the two antennas with and without the quartz-resonator is shown in Figure 3.7(b). The spacing between the two antennas is 0.5λ at 122 GHz. The mutual coupling between the antennas were simulated using the 3-D electromagnetic simulator, CST Microwave Studio, and is <-25 dB at 120 -123 GHz. A 3D- model including the wire-bonds and the QFN package is used for simulation, as shown in

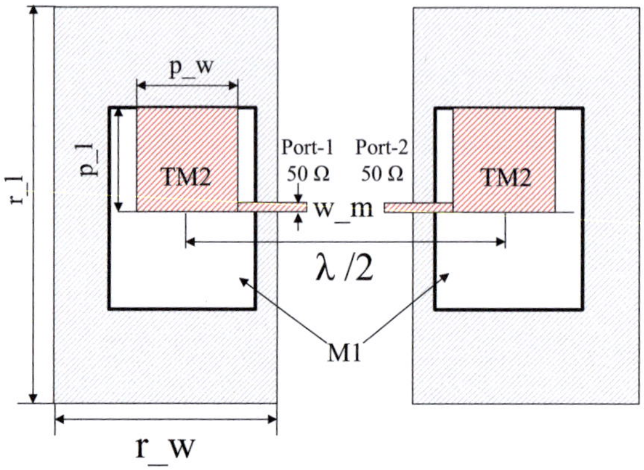

Figure 3.4: On-chip antenna layout

Figure 3.5. The simulated return loss and coupling of the antennas is depicted in Figure 3.7(a).

Label	Description	Dimension
p_l	on-chip patch length	290 μm
p_w	on-chip patch width	160 μm
w_m	micrsotrip feed line width	16 μm
r_l	quartz resonator length	520 μm
r_w	quartz resonator width	250 μm
ϵ_r-silicon	Dielectric constant of silicon	11.3
ϵ_r-glass	Dielectric constant of glass	3.77

Table 3.2: Dimension and parameters of the on-chip antenna

The realized radar transceiver is packaged using a wire-bonding technology in a 5 mm × 5 mm open cavity QFN package, as shown in Figure 3.6. As all millimeter wave signals are routed inside the package, only direct current (DC) and signals with a frequency of less than 3.8 GHz need to be conducted through the package leads.

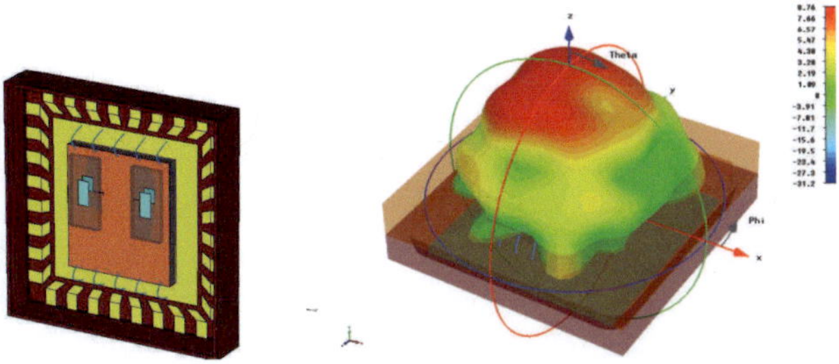

(a) 3D antenna simulation model (b) 3D radiation pattern

Figure 3.5: Simulation model and 3D far-field diagram of the two on-chip antennas.

Figure 3.6: Photo of the SMD radar in 5mm X 5mm QFN package.

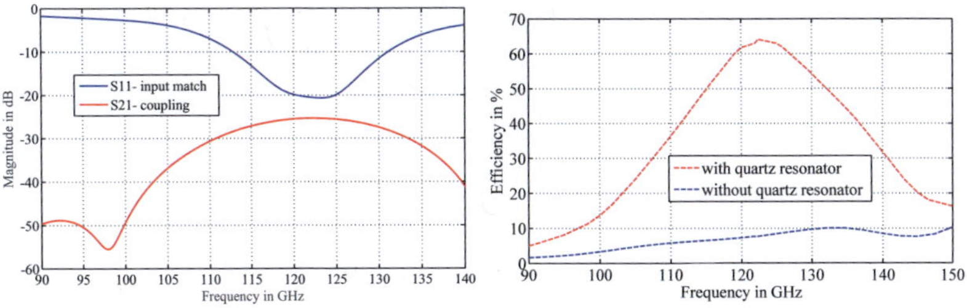

(a) S- parameter simulatin (b) Antenna efficiency

Figure 3.7: Simulated S-parameter of the two on-chip antennas: S11 is input match and S21 is coupling of the two antennas. Simulation results of antenna efficiency with and without quartz glass.

3.2.1 Antenna pattern measurements

The radiation pattern measurements are performed in an anechoic chamber, by rotating the radar sensor board to scan the half-sphere around the measuring antenna. In this measurement scenario the radar sensor is used as a transmitter and a standard gain horn antenna as a receiver in the far-field region. The antenna's measured and simulated radiation patterns are shown in Figure 3.9(a) and Figure 3.11, which show good agreement. This term represents the result of adding up all the radiation contributions of the two antenna elements. The theoretical and measured radiation pattern at 122 GHz exhibits a quite good agreement in E- and H-plane.

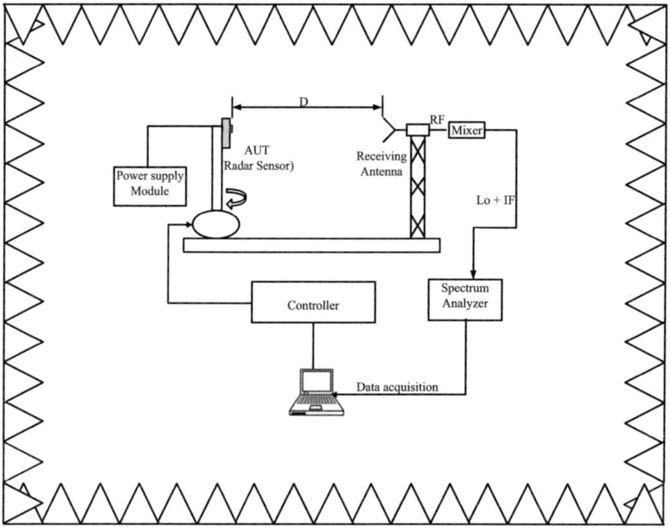

Figure 3.8: Antenna measurement set-up in anechoic chamber.

The radar sensor AUT (Antenna-Under-Test) is used as a transmitter and is mounted on a rotating table, which can be controlled by a position control unit. The receiving antenna, which is a standard horn antenna, is connected to an external mixer for down-conversion and to a spectrum analyzer. The distance, D, between the two antennas is adjusted to be more than $D > 2d^2/\lambda \approx 0.48\ m$

to make sure that the measurement is in the far field. To automate the entire measurement sequence a matlab program is used, which controls both the stepped rotation of the radar sensor and the spectrum analyzer simultaneously. The received RF signal is converted to IF by using the external mixer. The spectrum analyzer is adjusted in external-mixer mode, so that it sends a reference LO signal to the external-mixer. The theoretical radiation pattern in

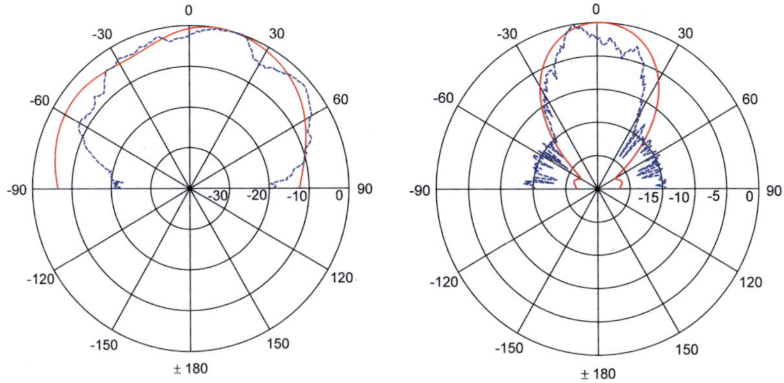

(a) E-plane radiation pattern (b) H-plane radiation pattern′

Figure 3.9: Measured (- -) and simulated (–) E- and H-plane radiation patterns in dB. The values are normalized to the maximum.

the E-plane is nearly omni directional since the antenna is simulated with an infinite ground plane. Practically, the packaged antenna is soldered on a test board with a limited ground plane and several wire-bond connectors affecting the radiation at elevation angle close to the horizon.

Figure 3.10: A measurement set-up for antenna measurement and sensor performance characterization. A dielectric lense deployed above the RF-board.

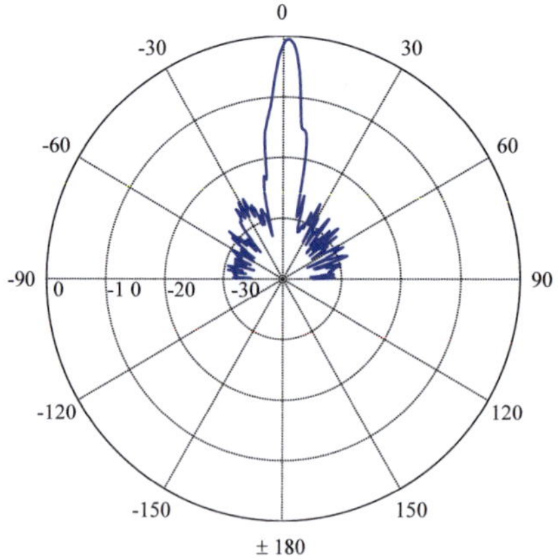

Figure 3.11: Measured radiation pattern with dielectric lense is employed above the MMIC.

3.3 Sensor hardware integration

Figure 3.12 shows the block diagram of the complete radar sensor module. In order to facilitate easier sensor module development, the sensor was split into two separate components: a baseband board and a radar board. The radar board includes the 122 GHz radar transceiver, a PLL chip for frequency control and modulation, and local power supply. The baseband board contains the

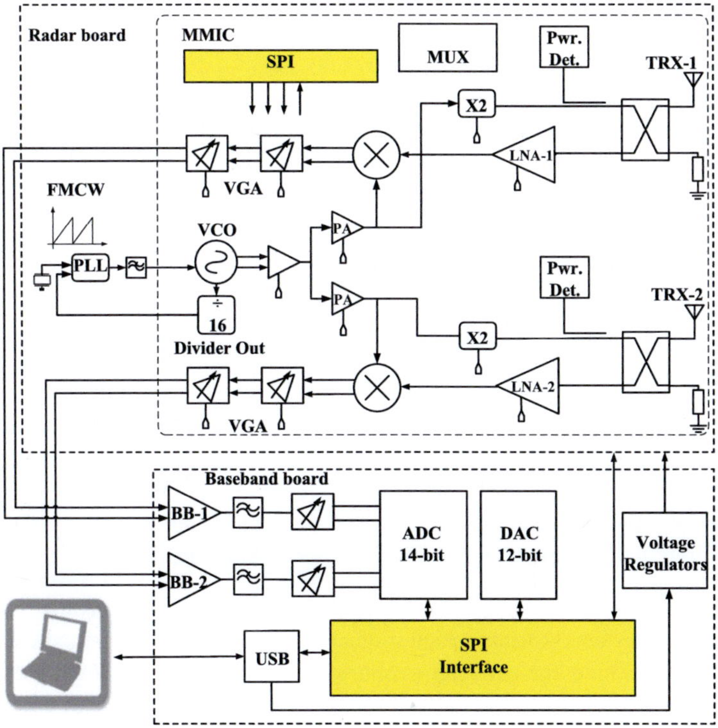

Figure 3.12: Block diagram of the radar module, which contains the radar board with a frequency control unit, and the base-band signal conditioning circuitry

analog and signal conditioning circuits, as well as a SPI interface, for digital communication. The baseband board is mounted at the back of the radar board using standard dual-row pin headers. To allow a flexible signal processing, the digitized base band signals are transferred to a PC via USB-Interface.

3.3.1 RF-Frontend and PLL circuitry

The RF-frontend board comprises of the integrated SMD chip, PLL circuitry, a low-pass filter and voltage regulator to supply the radar chip. The received base-band signals is routed to the base-band board mounted at the back side of the module via header connectors.

To operate the radar transceiver in FMCW mode a PLL circuit which is capable of modulation is required. A basic PLL circuitry contains a PLL chip a phase-frequency detector (PFD), charge-pump (CP), loop filter, and a prescaler as a first stage. The internal divider of the PLL chip can be used to prescale the input RF- and reference signals as needed.

Integrated PLL: The radar module consists of a commercially available fractional-n PLL, ADF4158, with direct modulation and waveform generation capability. It consists of a low noise digital PFD, a precision charge pump, and a programmable reference divider. For FMCW modulation, the frequency ramp is generated by altering the fractional divider value over time.

Divider signal: The VCO includes a divide-by-16 prescaler to directly connect to the input of the external PLL chip. The 3.8 GHz RF is subsequently prescaled in the PLL chip prior to the input of the PFD. A reference frequency of 30 MHz is provided to the PLL.

Loop-filter: The PLL loop-filter is needed to remove any unwanted high frequency components which might pass out of the PFD and appear in the VCO tune line and later as spurious signals. A passive loop-filter is designed, since the out put voltage of the charge-pump is high enough to control the VCO and cover the tuning band width of the VCO. The PLL is configured with loop-filter bandwidth is set to 25 KHz with a tuning sensitivity of 5MHz/V.

Figure 3.13 shows the PCB of the radar board. A close-up view of the sensor that is soldered onto the PCB and the quartz resonator is shown in Figure 3.13. To protect the IC and the bond wires from outside influences, a protecting

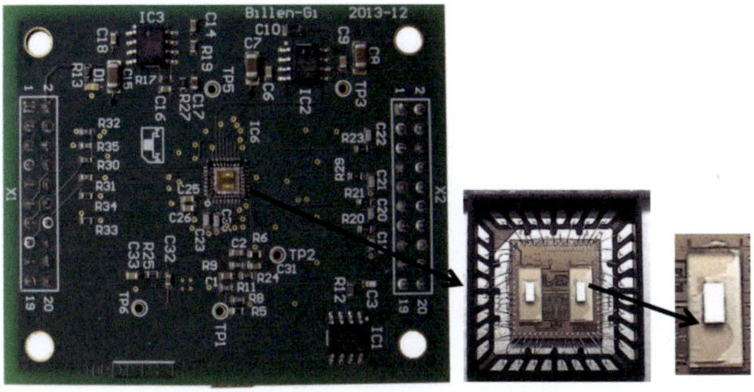

Figure 3.13: SMD radar sensor mounted on a low-cost PCB, 50 mm × 60 mm.

Figure 3.14: Sensor covered with a protecting cap.

cap with a thickness of half the guided wave length at the center frequency of 122.5 GHz is designed, as shown in Figure 3.14. As the reflected waves at the two interfaces air-dielectric and dielectric-air effectively cancel out, the cap has a marginal effect on the antenna performance.

3.3.2 Base-band circuit

The received base-band signals are in the first step filtered and amplified in analog domain. This step of signal pre-conditioning is necessary to adapt the dynamic of the received signal to the ADC and match the mixer impedance to the input of the ADC. Furthermore, the base-band board consists of an analog receiver chain with an instrumentation amplifier, a two-channel ADC to digitize the receive signals, auxiliary ADC and DAC. In order to allow software-based adjustment of the input stage DC offset and ADC common mode voltage, and to provide additional analog control signals to the radar board, an additional 12-bit DAC was provided. The ADC, DAC as well as the radar MMIC are controlled by a micro-controller board which is connected at the back side of the base-band board and the PC via a high speed USB connection. Using

Parameters	**Value**
Modulation bandwidth	Δf = 1500 MHz
Modulation sweep time	T_{sweep} = 10 ms
ADC dynamic range	73 dB
Sampling rate	f_s = 145 KHz
ADC input impedance	2KΩ
ADC input voltage range	0 - 3.3V
Number of channels	2
Maximum distance	R_{max}=5 m

Table 3.3: Key baseband parameters and configuration of the radar prototype.

the maximum specified measurement distance of R_{max} = 5 m, and the corresponding FMCW operation parameters for sweep time T_{sweep} = 10 ms and a modulation bandwidth of $\Delta f = 1500$ MHz, the maximum beat frequency a the receiver can be computed to 5 KHz, from Equation 2.20. Consequently, a VGA with differential inputs, which have gain only in the first few KHz was selected. To filter out high frequency components that may have coupled into the analog input, a simple RC-filter is implemented prior to the amplifiers. The parameters configuration of the base-band board is listed in Table 3.3.

3.4 Signal processing chain

In order to analyze the estimation of performance with respect to the final quantity distance, the raw baseband data are processed in further signal processing steps. The subsequent computation steps of the distance and angle of a target are illustrated in Figure 3.15. The first step of the signal processing is to select a valid region of the captured signal, in which the PLL is locked. Further, the mixers in the radar transceiver exhibit a frequency dependent DC-offset. In order to remove this signal trend, a linear regression is performed. The de-trend algorithm eliminates possible DC-offset voltages as well as linear slopes introduced by the frequency modulation. As an optional step, subsequent baseband signal frames could be collected to perform averaging, which increases signal to noise ratio.

A Hanning weighting function and optionally zero-padding is used prior to the FFT to compute the distance of the received spectrum. In order to estimate treshold levels for peak detection, the logarithm of the magnitude of the frequency domain signal is calculated.

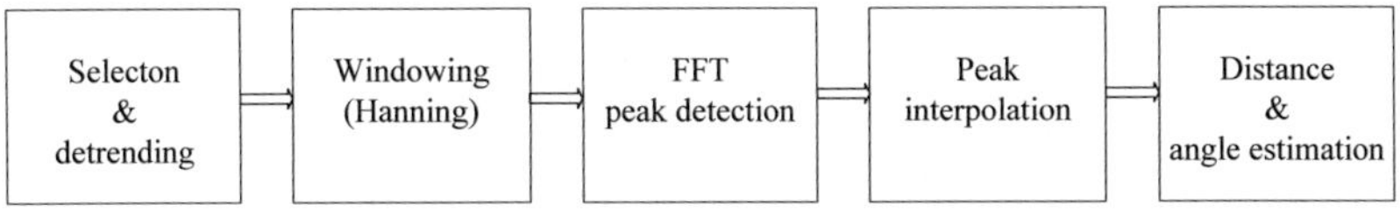

Figure 3.15: Flow chart illustrating the signal processing.

3.5 Performance and experimental results

Different Measurements of a single-target scenario with a corner reflector have been performed in an anechoic chamber to evaluate the achievable range, measurement accuracy and angular estimation performance of the transceiver chip. A dielectric lens above the RF board provided a narrow elevation beam and an increased relative antenna gain, resulting in a better signal-to-noise ratio (SNR). While range measurements have been taken both with and without the dielectric lens, angular measurements have only been taken without the lens, as the shift of the two antenna positions against the focus point of the lens leads to divergent influences on the received phase.

The measurement setup consists of a linear drive which can be used to precisely displace a corner reflector in a distance between 0 m- 1 m with a precision of about 12.5 μm. For all measurements, an FMCW modulation with start frequency $f_c = 121.5\,\text{GHz}$, bandwidth $\Delta f = 1500\,\text{MHz}$, and ramp duration $T_s = 10\,\text{ms}$ was used. The sampling rate of the ADC was set to 145 kHz. Digital signal processing, controlling the parameters of the transceiver via the SPI interface, and read-in of the measurement data was performed on a standard PC.

3.5.1 Range measurement

The first range measurement was performed with a dielectric lens above the RF board and a corner reflector as target, that was moved on a linear drive from $1\,\mathrm{m}$ to $2\,\mathrm{m}$ in steps of $1\,\mathrm{cm}$. At each position, 20 measurements have been taken to assess the reproducibility of the range measurements. Figure 3.17(a) shows the resulting estimation error dependent on the actual distance. The systematic error is estimated as the mean of the 20 measurements for each range. For this specific measurement example, the absolute deviation is $\pm\,3\mathrm{mm}$. A

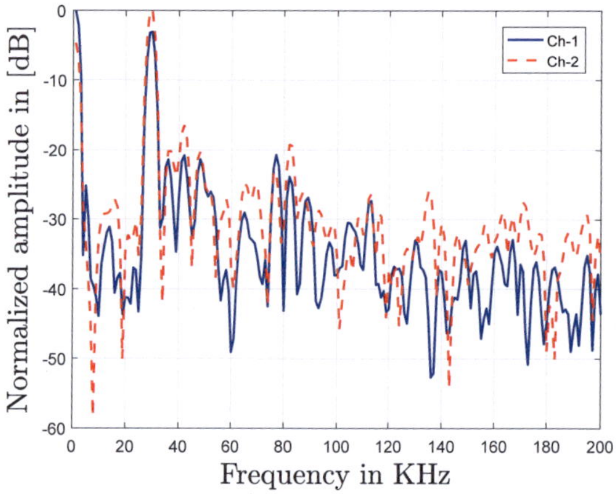

Figure 3.16: Range measurement of a single target.

second range measurement was performed without the dielectric lens and with an adapted baseband signal filtering. Here, the target was moved from $0.5\,\mathrm{m}$ to $3\,\mathrm{m}$ in steps of $0.5\,\mathrm{m}$. Figure 3.17(b) shows the results of this measurement series, where the error bars represent the standard deviation of the random error. Note that the lower RF power due to the missing lens required a slight modification of the analog baseband processing. The baseband signals were AC-coupled, to minimize the effect of DC offset due to the crosstalk between transmit and receive channels.

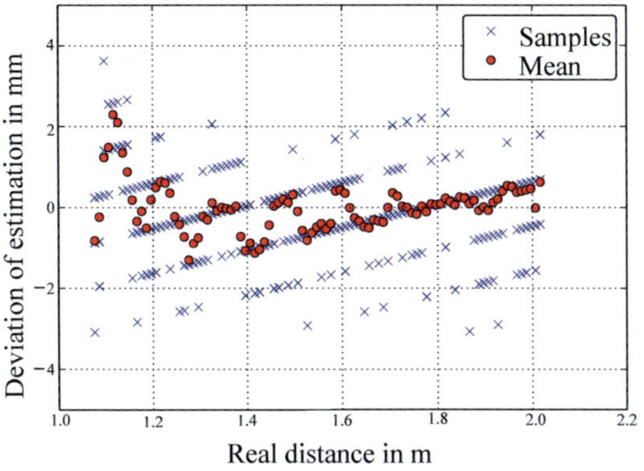

(a) Measured deviation values

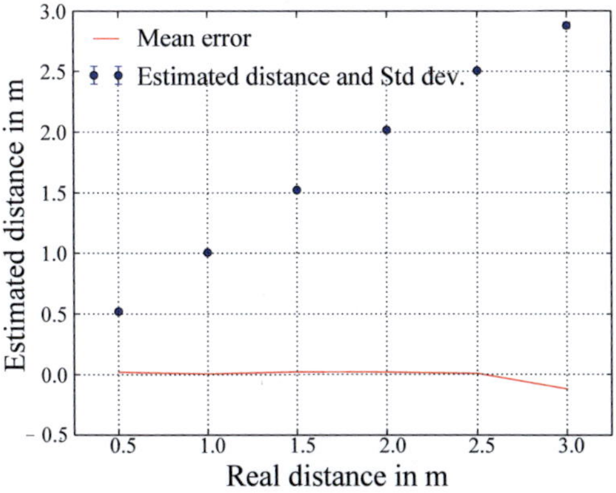

(b) Estimated distance and mean error value.

Figure 3.17: Distance estimation for single target corner refector at a distance between 0.5 and 3.0 m.

3.5.2 Azimuth angle estimation

For the evaluation of the angular estimation performance, a measurement setup with a stationary corner reflector and the radar sensor mounted on a rotating table is used. In this measurement scenario, no dielectric lens is deployed as it results in a very narrow antenna beam for angular measurements.

The radar sensor is rotated from angles of $-90°$ to $+90°$ in steps of $5°$. The angle-of-arrival is calculated by phase difference measurement of the receive channels, as described in section 2.1. In every position, 20 measurements have been taken to assess both the systematic and the random error, represented in the mean and standard deviation of the measurements in one specific position. This distinction is crucial, as systematic errors can be removed by calibration, while random errors remain.

In Figure 3.18(a), the results of the azimuth angle estimation using the conventional beam forming algorithm of the two channels is depicted. Here, the lengths of the error bars show the standard deviation of the random error. At wider angular position of the radar sensor, both the mean and random error of the angular estimation increase due to the decreasing SNR. This behavior results from the strongly decreased SNR at larger angles in accordance to the two-way antenna diagram in Figure 3.18(b).

3.5.3 Two-way antenna pattern

Additionally, the two-way antenna diagram was calculated from the received power in the same single-target scenario over different angles. The mean output power of the receive mixer of a single receive channel is used as an estimate for the received power in conjunction with a reflecting target at various angles. This estimates the two-way beam pattern of the antenna, as shown in Figure 3.18(b). The antenna characteristic determined using this method represents the combination of receive and transmit antenna beam pattern. For the two-way antenna diagram, a more distinct asymmetry than in the E-plane one-way antenna diagram in section 3.2.1 can be observed.

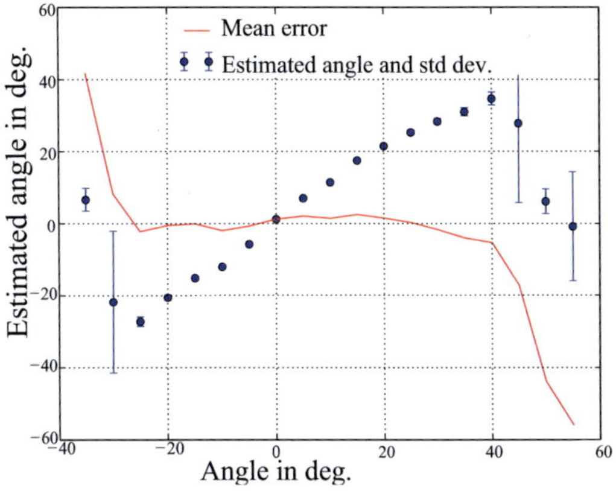

(a) Estimated angular position

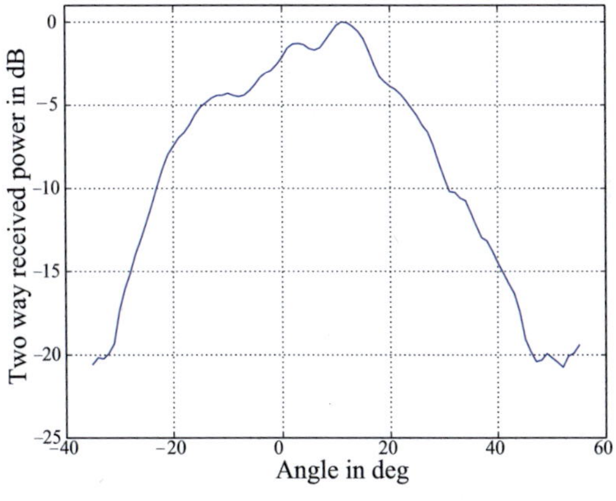

(b) Two-way antenna pattern

Figure 3.18: Estimated angular position and standard deviations determined from 20 measurements. The two-way antenna pattern obtained from the target response. The measurement setup shown in Figure 3.10 is used without the dielectric lens.

3.6 Summary

A fully integrated, dual-channel, single chip transceiver has been realized for operation at D-band frequencies. The transceiver is based on 130 nm SiGe-BiCMOS technology with two integrated on-chip antennas. The SMD radar requires no external millimeter wave connections, thus enabling the assembly in a standard SMD line. Due to the two receiving channels, angular measurement in azimuth direction is possible. The specifications and system concepts of the proposed D-band radar front-end and baseband analog signal processing blocks are presented. These results demonstrate the potential of compact D-band radar sensor for a consumer application.

A performance limitation is observed on the radar system developed in this chapter, as it is not capable of measuring short range distances below 0.2 m. This is due to the homodyne architecture of the system, suffering from DC-offset[1] issues which mask short range targets with beat frequencies near to DC.

[1] A detail description of DC-offsets in FMCW radar systems and a case-study are analyzed in Sections 4.1 and 4.7 respectively.

4 Novel heterodyne radar system architecture

The performance of radar sensors depends highly on the system design, architecture and the characteristics of components integrated on the system. In Chapter 2, heterodyne and homodyne radar architecture variants of radar receivers were described. Conventionally, most of the literatures on mm-wave radar sensor systems have been focusing on homodyne architectures as the ones described in Chapter 3. However, homodyne architectures have a DC-offset problem, which limits the performance of the sensor in the near range, compared to the heterodyne architectures where the IF signal is at non-zero frequency. A radar sensor with a homodyne receiver architecture is detailed in Chapter 3.

Recently there has been a technology moving towards using heterodyne architectures instead. Although the hardware required for realizing such systems is slightly more complex, this configuration has the potential to remove problems related to DC-offset errors and accurate measurements of very short distance. In this chapter a system architecture for heterodyne system is studied and within this PhD work a demonstrator is built to verify the advantage of heterodyne systems for short range sensing.

4.1 DC-offset thematic in radar systems

In FMCW radar system, DC-offsets occur due to several factors, including DC-offsets inherent to the circuits, self-mixing, unwanted reflection from stationary near-by objects, and DC values that are due to signal leakages in system, antenna coupling etc. These effects generate a beat frequency close to zero, which is why it is often termed as a DC-offset issue, as shown in Figure 4.1. Several studies in recent development have been focusing on the elimination of DC-offset in radar systems. Methodologies for leakage cancellation are shown in [LWPS06, Sto92]. A study and analysis on the impact of non-idealities in FMCW radar systems, are studied in [TEV13, WSJ06]. The primary consequence of DC-offset in radar systems is, a performance degradation and failure of object detections in short range.

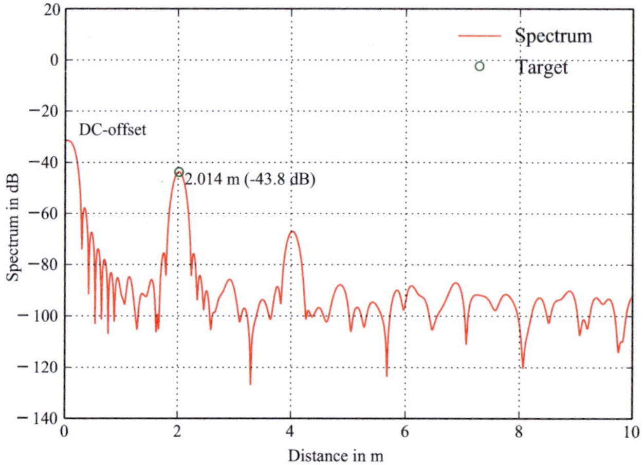

Figure 4.1: An example of a DC-offset spectrum measured with the homodyne system described in chapter 3, which shows a DC-offset block which could mask targets up to nearly 30 cm.

DC-offset due to self-mixing

DC-offset due to self-mixing occurs when the RF output signal mixes with the LO, either through substrate coupling or from reflections at the antenna connection. Additionally, due to finite isolation between the ports of the constituent mixer signal leakages between the LO and RF ports occur. The leakage of the RF signal to the LO port is typically negligible due to the small RF power as compared to that of the LO signal, as shown in Figure 4.2(a). However, the leakage of the LO signal onto the RF port can be significant. This LO-leaking signal mixes with the original LO signal to produce a signal which results in a DC component. This DC offset is undesirable, as it limits the dynamic range of the receiver without adding information about the target. The LO may be conducted or radiated through an unintended path to the mixer's RF input port, thus effectively mixing with itself, producing an unwanted DC component at the mixer output. This LO leakage may reach the LNA input,

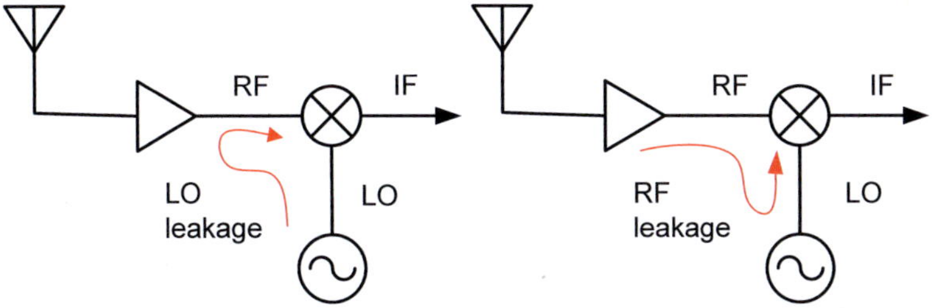

(a) Self-mixing due to LO leakage (b) Self-mixing due to RF leakage

Figure 4.2: Self mixing due to different leakage paths in radar receiver, [SR05].

producing an even stronger result. This effect presents a high barrier against the integration of LO, mixer and LNA on a single silicon substrate, where numerous mechanisms can contribute to poor isolation.

Conversely, a strong interference signal, once amplified by the LNA, could find a path to the LO-input port of the mixer, thus once again producing self-mixing, as in Figure 4.2(b). Due to their non-ideal isolation to the antenna, some amount of LO power will be conducted through the mixer and LNA. The leakage of LO or RF signals to the opposite mixer port is not the only way in which unwanted DC can be produced.

DC offset due to reflection from stationary targets in short range

Another source of DC-offsets, is a reflection from targets with a small round-trip delay time. In FMCW radar systems, a fixed target in front of the radar antenna, for example, a fixture or radome, whose intermediate frequency impact is unwanted and possibly interferes with the beat frequency of the other targets will result in a failure detection. This kind of the transmitter power leaking into the receiver raises significant problems for an FMCW system that must receive echoes while it is transmitting. These problems include a limit to the maximum transmit power and a requirement for careful matching of the antenna and a coupler to minimize the magnitude of the reflections.

Figure 4.3 presents a simplified schematic of DC-offset sources in FMCW-radar on the basis of a monostatic approach. The signal for the local oscillator LO serves both as a feeding signal for the mixer and as transmitting signal of the radar. In addition to direct coupling of the LO-leakage, Γ_{LO}, in the mixer core, reflections of the transmitting signal from a stationary target near to the antenna, are responsible for a DC-offset. Becuase of the short propagation path the frequency in the FMCW-ramp has not changed when the reflection return to the RF-port. If the reflected signal is too large it can bias the mixer out of its operation region into saturation and degraded performance results. For a monostatic system bond connection, or impedance mismatch of the antenna, Γ_{Ant}, are dominant. Whereas, in a bistatic radar these reflections do not apply due to the fact that the transmitting and receiving antennas are separated.

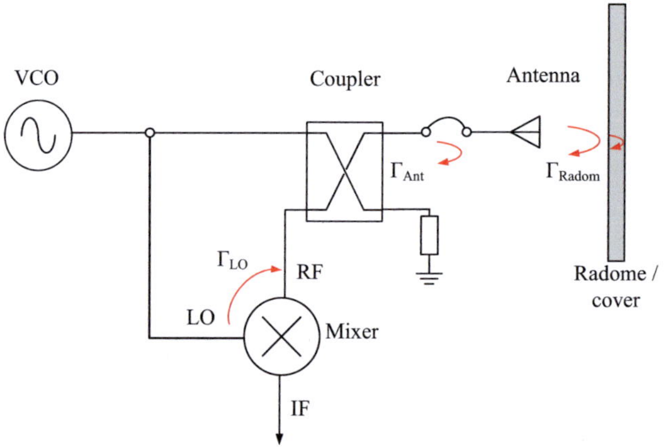

Figure 4.3: DC-offset sources due to LO leakage and short range leakage in FMCW radar system: where Γ_{LO} = LO leakage, Γ_{Ant} = antenna or connector mismatch, and Γ_{Radome} = a short range reflection from radome or cover.

In addition to these two described effects, the leakage signal also introduces additional noise into the receiver over and above the thermal noise level. This additional noise is due to the phase noise of the transmitter, which will be described in section 4.3.1.

Impact of DC-offset on radar systems

The RF-LO leakage will affect in the first place the performance of the mixer, which mainly depends on the amplitude level of the leakage signal. The resulted DC-offset at the IF port will saturate the subsequent components, such as base band amplifier (BBA) and Analog-digital-converter (ADC). The impacts of DC-offset on the radar system are:

- Limitations in near range detections due to masking of objects with the DC-offset near to $f_{beat} = 0$
- Performance degradation of mixer circuit, due to overloading and saturation of the mixer components.
- Decrease of SNR in the radar system due to increase in noise
- Influence on the valuable dynamic range of ADC device

Sofar, to use AC coupling, high pass filtering, or DC feedback loop are the most popular and common solutions , as described in [HTC96], [PSL$^+$07]. These methods can be effective in eliminating the static DC-offset, but cannot cancel the dynamic DC-offset which is varying fast over time and system temperature. Elimination of DC-offset by using AC coupling is more effective when the transmit signal is constant. However, in FMCW radar systems, where the frequency is modulated over time, elimination of DC-offset is not possible, which in turn results in a performance degradation of the sensor and failure of object detection in near range.

4.2 Heterodyne radar systems at mm-wave frequency

The working principle of a heterodyne radar architecture is described in section 2.3.1. The drawbacks of homodyne radar system can be solved by heterodyne architecture, achieving a frequency separation between the RF and LO ports. In [DSIP13], frequency separation is achieved by multiplying one source with different factors. Another approach is to mix the sweeping source with two different fixed sources and then multiply with the same factor to acquire the separation [CDC$^+$08, CDL$^+$11]. In [FKS$^+$11], a 77 GHz heterodyne FMCW radar with offset PLL frequency is proposed. An alternative method is to use two different VCOs with a frequency offset. However, this method is sensitive to the uncorrelation of phase noises which is detailed in section 4.3.1.

In most of heterodyne FMCW systems a cancellation of correlated phase noise components being present in the transmit and downconverter part of the system is implemented. Such a cancellation can be achieved by using a reference path with either additional mixer stages [CDC+08, DSIP13] or by digitizing the reference signal in order to cancel correlated phase noise by digital signal processing techniques [MS01, FKS+11]. Furthermore, multi-VCO based concepts can be combined with frequency multipliers operating at differing multiplication factors to realize frequency offsets [CDL+11].

4.3 Single chip 122 GHz heterodyne radar architecture

A single chip heterodye radar sensor at 122 GHz is realized and radar system prototypes are built, within the frame of this PhD work. The mm-wave frontend, including the on-chip antenna, is based on an integrated circuit realized in a 0.13-μm SiGe-BiCMOS technology with f_T= 230 GHz and f_{max} = 280 GHz [RHW+10]. The circuit blocks of the transceiver has been designed by University of Toronto and is detailed in [SHBV12] and [Sar13], so only a brief introduction with an emphasis to the system aspects will be given. The radar system design, built-up of the system prototypes, and all performance measurement and analysis are parts of this PhD work.

4.3.1 System architecture

The block diagram of the 122 GHz transceiver chip is shown in Figure 4.4. It is realized as a heterodyne system, having separate transmit and receive oscillators, all integrated in a single chip. On the transmit side, the oscillator signal is split and one part is fed into a power amplifier with a maximum output power of 0 dBm. Using a 6 dB-coupler to separate receive and transmit signals, the transmit signal is then radiated by the on-chip antenna.

At the receiver side, the receive oscillator drives two mixers and a divide-by-64-chain. The divider chain allows the frequency stabilization of the receive oscillator, whereas the two mixers provide IF signals by mixing the receive signal or the signal from the transmit oscillator directly. The mixers provide

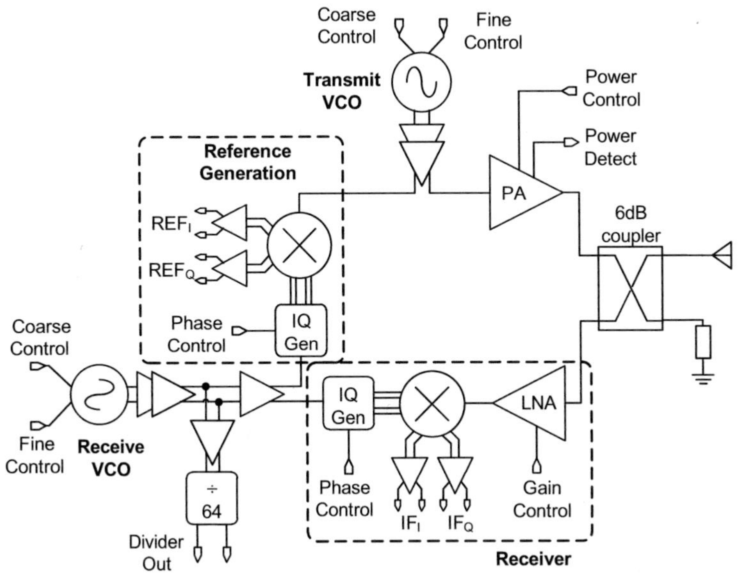

Figure 4.4: Block diagram of a heterodyne radar system with on-chip antenna, [Sar13].

IQ signals for the reference and receive IF signals. Due to the analog nature of the IQ generation, no perfect 90° phase difference can be obtained over the whole 1 GHz operation bandwidth, so a phase control input is provided to adjust the phase difference.

From a system design point of view, the implemented architecture provides a great deal of freedom at the sensor operation side. In continuous wave(CW) operation, by providing IQ receive signals, at an IF the 1/f noise issues of direct conversion systems can be avoided. In frequency-modulated continuous-wave(FMCW) operation, the DC offset issues can be mitigated to a large extend, avoiding problems like mixer bias changes caused by large reflections.

As already pointed out above a heterodyne radar system require a two signal sources that possibly shift the baseband signal to a non-zero IF frequency. The two signal sources are used to generate a tunable frequency f_{TX}(t) and f_{RX}(t) with t denoting the continuous time, by certain constant offset frequency generated by the two oscillators. The signal generated by oscillator 1 is transmitted by the antenna and received back.

Phase noise and range correlation

The phase purity of a radar signal in FMCW radar systems is a crucial design criterion. Particularly due to the high frequencies in today's radar systems, the phase noise becomes a dominant noise source affecting the detection accuracy and sensitivity which can be limiting factor to the system. In direct-conversion radar systems, as described in chapter 3, the same source is used for the transmitted signal and the local oscillator signal in the receiver, which means the received signal is a time-delayed version of the transmitted signal. Therefore, the phase noise of the received signal is correlated with that of the local oscillator, with the level of correlation dependent on the time delay between the two signals. When the two signals are mixed, the correlated portion of the phase noise effectively cancels, leaving a residual phase noise spectrum at baseband that is far below the phase noise spectrum at RF.

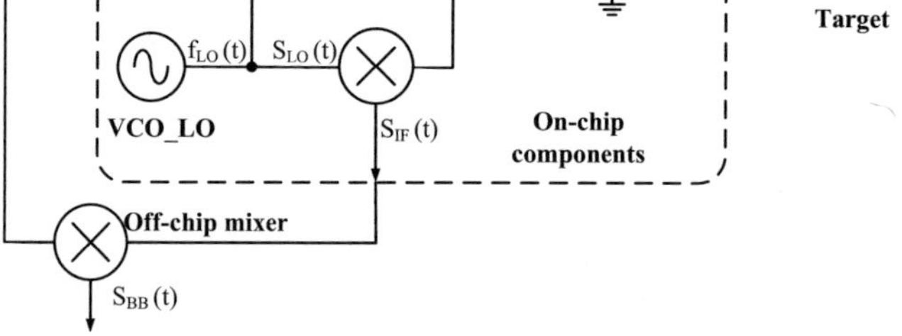

Figure 4.5: A simplified signal flow diagram of the heterodyne radar system described in section 4.3. The base band signal is obtained by a further down conversion of the IF and REF signal, thus an additional off-chip mixer is necessary.

Consider the heterodyne radar system depicted in Figure 4.5, which is the simplified representation of the system shown in Figure 4.4 with phase noise included in both oscillators. Unlike in homodyne system, the IF signal in heterodyne system is generated by the multiplication of two signal sources, f_{TX} and f_{RX}, in the down-converter. As a result, the phase noise of the two signal sources is a significant noise contributor on the whole system. However, as shown below, near range radars have an inherent advantage in dealing with phase noise compared to long-range radars.

The signal generated by the transmitter and the LO-VCO is given by:

$$S_{TX}(t) = A_{TX}cos(2\pi f_{TX}t + \phi_{TX}(t)) \tag{4.1}$$

$$S_{LO}(t) = A_{LO}cos(2\pi f_{LO}t + \phi_{LO}(t)) \tag{4.2}$$

where $\phi_{TX}(t)$ and $\phi_{LO}(t)$ is the phase noise signal generated by transmit VCO and the LO or the reference VCO respectively. The reference output signal becomes:

$$S_{REF}(t) = A_{REF}cos(2\pi(f_{LO} - f_{TX})t + \underbrace{\phi_{LO}(t) - \phi_{TX}(t))}_{\text{added phase noise term}} \tag{4.3}$$

The received signal at the input of the receiver mixer after the round-trip delay τ_D is:

$$S_{RX}(t) = A_{RX}cos(2\pi\ f_{TX}(t - \tau_D) + \phi_{TX}(t - \tau_D)) \tag{4.4}$$

which is given by $\tau_D = 2R/c$, where R is the range to the target, and c is the speed of light. After down conversion to IF,

$$S_{IF}(t) = A_{IF}cos(2\pi f_{LO}t + \phi_{LO}(t) - 2\pi\ f_{TX}(t - \tau_D) - \phi_{TX}(t - \tau_D)) \tag{4.5}$$

To obtain the base-band signal or the beat frequency, a further down-conversion of the IF and REF- signal is necessary, which is realized by using an off-chip mixer. The phase of $S_{IF}(t)$ and $S_{REF}(t)$ are then subtracted to calculate τ_D, yielding:

$$\Delta\,\phi = 2\pi\ f_{TX}\tau_D + \underbrace{\phi_{TX}(t) - \phi_{TX}(t - \tau_D)}_{\Delta\,\phi_n(t)} \tag{4.6}$$

where $\Delta\phi_n(t)$ is the error due to phase noise of the oscillator. If the time delay, τ_D, is small, the subtraction of the phase noise signal from the delay replica leads to a noise cancellation effect, known as phase noise correlation or range correlation, [BB93].

If the single-sideband power spectral density of $\phi_{TX}(t)$ is S_ϕ, the spectral density of $\Delta\phi_n(t)$ is given by, [BB93]:

$$S_{\Delta\phi} = S_\phi \cdot 4sin^2(\pi\tau_D f) = S_\phi \left[4sin^2\left(2\,\pi\frac{R\Delta\,f}{c}\right)\right] \tag{4.7}$$

where f is the frequency offset from the carrier. Therefore, range correlation acts on the noise close to the carrier and attenuates it. The correlation gain is typically high for a small value of the sinus function, i.e. for small value of τ_D and/or for small value of offset frequency Δf. For short ranges, where the time interval of the two-way range delay between transmission and reception is short, these phase errors remain correlated to some extent and tend to cancel. Figure 4.6, shows a simulation example of a phase noise correlation versus target range based on the equation 4.7. For closer ranges (smaller τ_D) a radar receiver cancels the phase noise at a rate of 20 dB/decade.

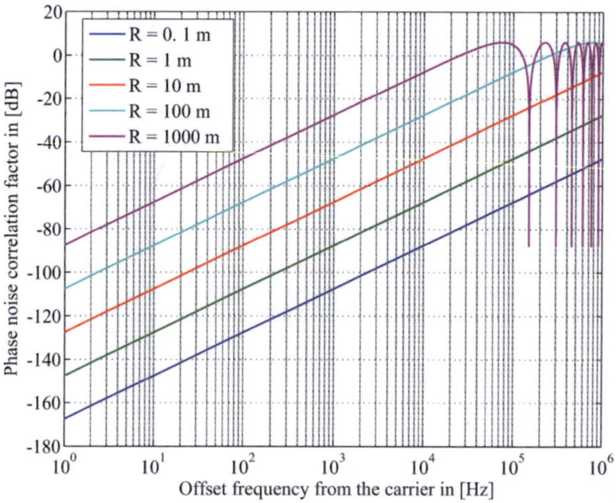

Figure 4.6: Phase noise correlation versus offset frequency for various target ranges.

For 122 GHz short range radar sensor, the two way delay between an object at a distance of 2 m is maximum τ_D = 13.3 ns, by shorter distance even less. This effect is beneficial because it reduces the effect of phase noise on short range, and it effectively attenuates the received phase noise power density spectrum.

Injection locking in heterodyne systems

The phenomenon of injection locking in analog and mixed-signal systems is the effect that can occur when an electronic oscillator is disturbed by a second oscillator operating at a nearby frequency. When the coupling is strong enough and the frequencies are near enough, the second oscillator can capture the first oscillator, causing it to have essentially identical frequency as the second. Considering the heterodyne radar system in Figure 4.4, the two on-chip VCOs may pull each other as a result of coupling through the substrate. The strength of coupling between the two VCOs is inversely proportional to their frequency difference and can lead to mutual synchronization when the frequency difference is small. Experimentally we have determined that this happens typically for a frequency difference $f_{IF}(t) = f_{TX}(t) - f_{RX}(t) \leq$ 600 MHz.

4.3.2 Integrated on-chip antenna

The on-chip antenna element was for the first time introduced in [HWGH10] at frequency of operation at 77 GHz. The same antenna concept is adapted for the 122 GHz application as in chapter 3.2. The on-chip antenna consists of two parts, where the first element is a primary radiator, integrated on-chip in the standard SiGe technology. The second element is a parasitic half-wavelength resonator, realized on a small quartz glass substrate glued on top of the chip, as can be seen in Figure 4.7. This half-wavelength structure couples the electromagnetic signal from the on-chip resonator. The main drawback of on-chip antenna are poor radiation efficiency and low gain. Without this parasitic resonator, the radiation efficiency of the antenna would be very poor. As shown in Figure 4.8 the integrated radar sensor requires no external mm-wave connection, thus it can be soldered in a low-cost RF-board. The RF subsection comprises mainly the packaged radar IC, supply voltages, and miniaturized coaxial connectors. A standard QFN 7×7 mm open cavity package with 48 pins was used to package the chip for easy mounting on a printed circuit board, as shown in Figure 4.7 and 4.8 respectively. The MMICs were bonded to the

Figure 4.7: 7mm x 7mm QFN packaged single chip radar sensor

Figure 4.8: SMD radar sensor mounted on the RF-board.

package frame using 25 μm diameter gold bond wire which is employed for the short length. Note that, all signals coming out of the chip are differential, in order to provide an interface to the baseband circuit that does not suffer from interference or noise. The IC circuit has a total power consumption of 900 mW from a 1.8 and 1.2 V supplies.

A dielectric lens is deployed above the RF-board to obtain a narrow elevation beam and an increased relative antenna gain. Antennas with dielectric lenses will focus most of their power into the dielectric side, and the field of view of the antenna system is easily tailored to the application by choosing a suitable diameter and and other design parameters of the lens. A photograph of the RF board including a suitable dielectric lens with holder is shown in Figure 4.9.

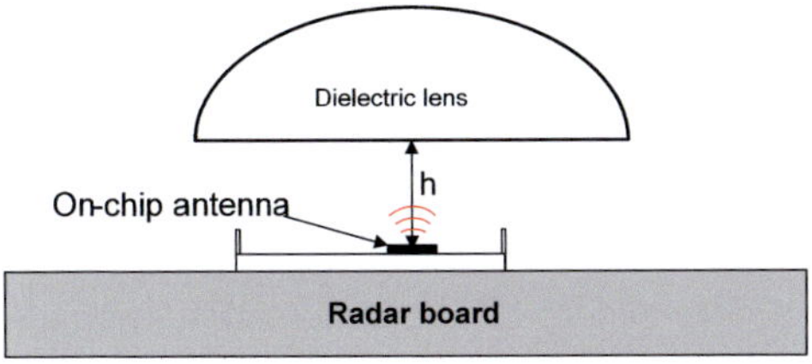

Figure 4.9: Radar sensor set-up with dielectric lens depicted on the top.

Antenna measurement

The radiation pattern measurements are performed in an anechoic chamber, by rotating the radar sensor board to scan the half-sphere around the measuring antenna. In this measurement scenario, the radar sensor is used as a transmitter and a standard gain horn antenna as a receiver in the far-field region. The same measurement set-up as described in Section 3.2.1 is used. The measured

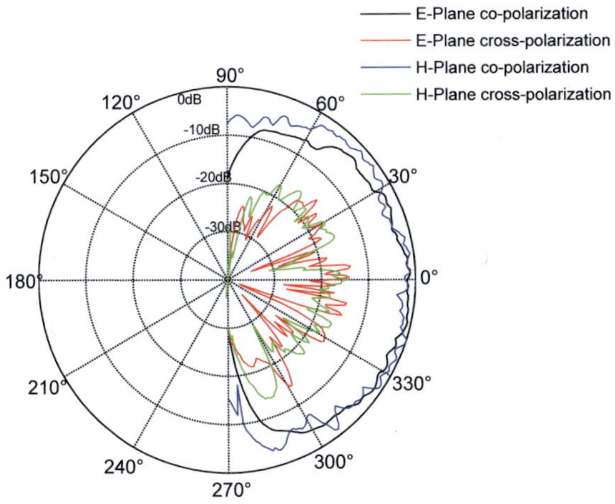

Figure 4.10: Measured antenna pattern in E- and H- plane, sensor without dielectric lens.

antenna pattern of the on-chip antenna is shown in Figure 4.10. It shows a very broad beam-width of around 150° in both planes. The measured radiation pattern of the antenna with and with out dielectric lens is shown in Figure 4.10

and 4.11, which shows a 20-dB gain improvement. Within the frame of this PhD work, further details of this antenna system is published in [GHS$^+$12].

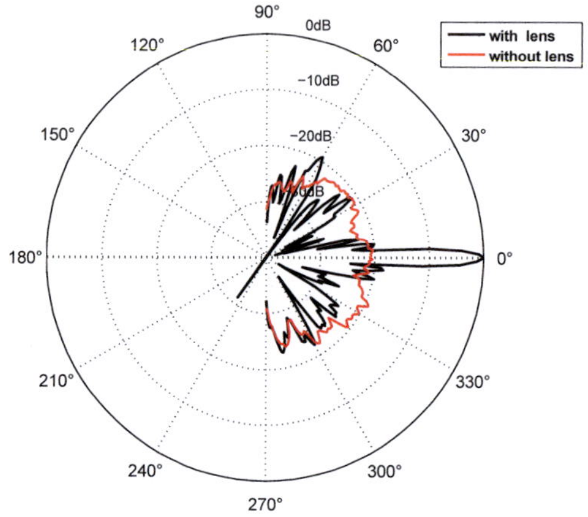

Figure 4.11: Comparison of measured antenna pattern with and without dielectric lens.

4.4 Radar system integration

Figure 4.12 shows the system block diagram of the radar sensor based on the radar system architecture described in section 4.3. In order to test the functionality of the whole transceiver, test boards were designed, which contains the necessary power supplies, several SMD components (PLLs, Mixers) and coaxial connectors for all relevant signals coming from the chip. A USRP-B210 SDR Software Defined Radio (SDR) platform is used to generate a reference signal for the PLL circuits and down sample the received RF signals for further processing.

Base band system consists of Universal Software Radio Peripheral (USRP) interface through USB3.0 is connected to the host PC. The signal of the PLL-TX and PLL-RX are generated from the GNU radio and transmitted though the transmission channel of the USRP. At the same time the IF and REF signals of

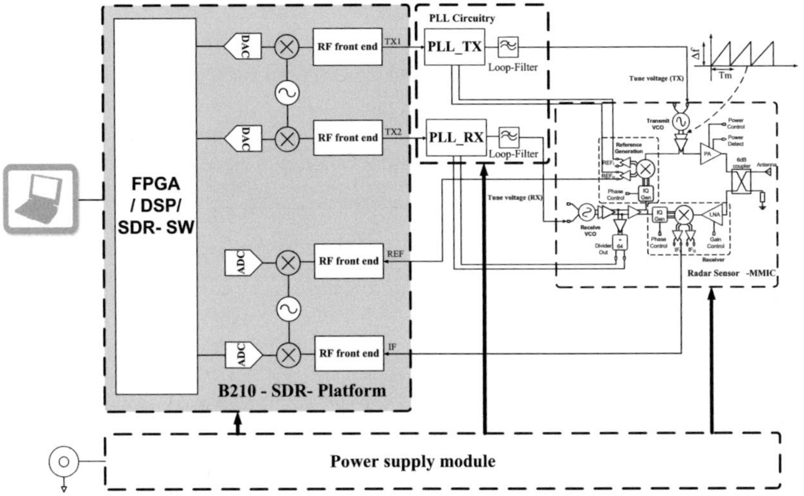

Figure 4.12: Block diagram of SDR based heterodyne radar system.

the radar are received by the two receiver channels for down-conversion. For the operation in FMCW mode the reference signal is modulated by continuously sweeping from start and stop frequency.

Overview of SDR

A low cost alternative to down convert an intermediate frequency is using an SDR module. The USRP-B210 is a single board version and consists of a spartan6 field programmable gate array (FPGA) processor, 2-ADC, 2-DAC and RF transceiver/RF- front-end. The sampling rate of the ADCs is 61.44 MSPS with a 12-bit resolution. The DAC is also capable of modulating to an intermediate frequency. The B210 has advantage of two transmitter and two receiver MIMO capabilities. In this configuration, an analog RF module is used to modulate signals between an intermediate frequency (or baseband) to the carrier frequency, f_c, and an FPGA is inserted between the ADCs and the processor.

Incorporating an RF module reduces the required sampling rate of the ADCs to $F_s \geq 2(F_i + 0.5\beta)$, when the intermediate frequency F_i is used, or $F_s \geq \beta$, when the baseband and a pair of ADCs (inphase and quadrature) are used. The FPGA can be used to assist with critical timing operations, up/down sampling, modulation, and matched filtering.

Type	Specification
Frequency band of operation	122- 123 GHz
Maximum transmit power	0 dBm
Bandwidth(Frequency sweep) in TX mode	1 MHz
Bandwidth	1024 MHz
Sweep time	10 ms
SDR IF frequency(F_i) RX mode	912 MHz
SDR LO frequency(F_i) in TX mode	114 MHz
Sampling rate (IF)	8 Msps
Decimation factor	80
Sampling rate (Base-band)	100 ksps
LPF cut-off frequency	100 kHz
Detection range (m)	0.1 - 5 m
Range resolution	15 cm
Antenna gain without dielectic lens	6 dBi
DC power consumption	900 mW

Table 4.1: Summary of realized radar sensor system level parameters

Unlike traditional fixed hardware implementations, SDR uses software configurable RF module which can be used to implement customized signal processing. SDR also provides a simple open source GNU Radio software that provides the means to develop real-time RF applications. The GNU radio package includes programming interfaces and run-time environments to implement the desired RF signal processing functionality, applications for creating signal flow graphs, and hardware drivers to access USRP devices. It uses the concept of signal processing blocks to achieve the required signal flow and processing functionality.

4.4.1 Frequency stabilization and modulation

Two commercially available PLL chips are connected to lock the transmit and receive VCOs, by setting the fine tuning voltage, as shown in Figure 4.12. The transmit VCO (VCO_{TX}) is stabilized by using the TX1 output of the B210 board as a reference. This is done by locking the reference IF output signals, which are already down-mixed in the chip. The receive VCO's (VCO_{RX}) divide-by-64 signal, which is in order of 1.89 GHz, is fed to PLL_{RX} to stabilize VCO_{RX}, by using the TX2 output of the B210 board as a reference. The VCO_{TX} operates with a small frequency offset to VCO_{RX}, thus the generated reference signal frequency is usually not greater than 1 GHz. As the maximum input frequency of the phase frequency detector (PFD) is 120 MHz, the divider signal is divided again on the PLL's N-counter, which is set to 32.

In principle, all basic radar modulation forms as FMCW, CW, or stepped-frequency modulation can be used as transmit signals of the transceiver modules. In FMCW mode of operation, ramp generation is done by sweeping the reference input frequency of the PLL_{TX}, which is set from the USRP board. The bandwidth and sweep time of the modulations parameters can be configured in the GNU radio companion. An FMCW signal with a time period of 10 ms and a BW of 1 MHz is generated which sweeps continuously from 114 MHz to 115 MHz is transmitted at a sampling rate of 8 Msps as TX1 to radar through USRP and CW at a frequency of 114 MHz transmitted as TX2.

4.4.2 IF data acquisition and signal processing

The base-band signal flow diagram and the implemented DSP blocks of the SDR-platform are shown in Figure 4.13. The received IF signals are further down-converted to base-band in analog domain at the initial stage of the receiver. The analog LO of the USRP board can be configured at the central frequency of F_I, by using the software based source in the GNU radio companion. Since the IF and REF frequencies are configured to 912 MHz. The sampling rate of the GNU radio is 8 Msps, which will be further reduced to 100 Ksps by using the decimation block. The received IF signals are further down-converted to base-band, by multiplying with the reference signals, in the digital domain. The mixed signal is first filtered to remove unwanted high frequencies in the processing stage. The filter output drives an amplification stage with selectable gain. After data acquisition, a further digital signal pro-

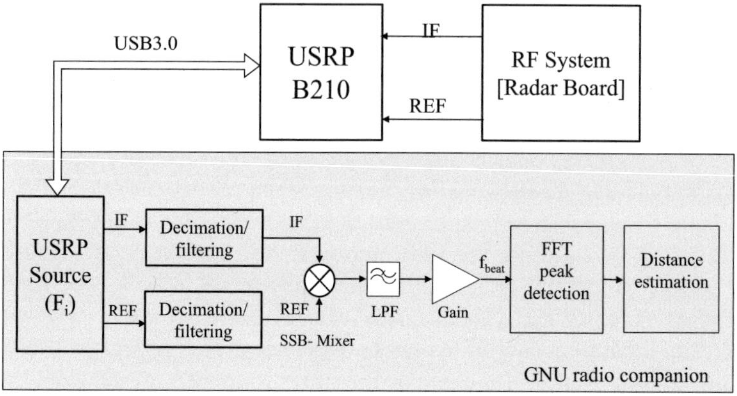

Figure 4.13: Receiver operation of the USRP board, which is implemented in the GNU radio companion.

cessing is implemented using a matlab program on the PC. This includes all necessary FMCW waveform settings and allows radar measurements of stationary targets. After signal selection and detrending, a Hanning weighting function is used prior to the FFT to compute the range of a target.

4.5 Performance and experimental results

The performance and functionality of the radar system was verified by a series of measurements using the system described in section 4.4. The primary motivation for building this SDR based mm-wave radar system was to show the performance of a heterodyne system for short range distance measurements. Furthermore, it allows to compare the measurement performance of both heterodyne and homodyne modes of operation under a comparable operating condition.

Several measurements of a single-target scenario with a corner reflector have been performed in an anechoic chamber to evaluate the achievable short range performance of the transceiver chip. At each position, 20 measurements have been taken to assess the reproducibility of the range measurements. Figure 4.14(a) to Figure 4.16(b) show the processed signal obtained from a corner reflector moved in a range of 0.1 m to 2.9 m, using a set of measurement

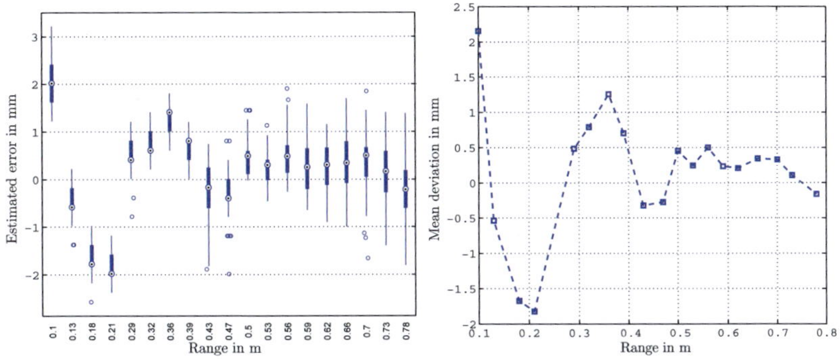

(a) Measured distance

(b) Measured mean value

Figure 4.14: Distance estimation for single target corner reflector from 0.1 m to 0.8 m.

position of a corner reflector moved on a linear drive, which is 1 m long. Due to the length limitation of the measurement set-up, 3 separate measurements are done by moving the linear-drive to the next range.

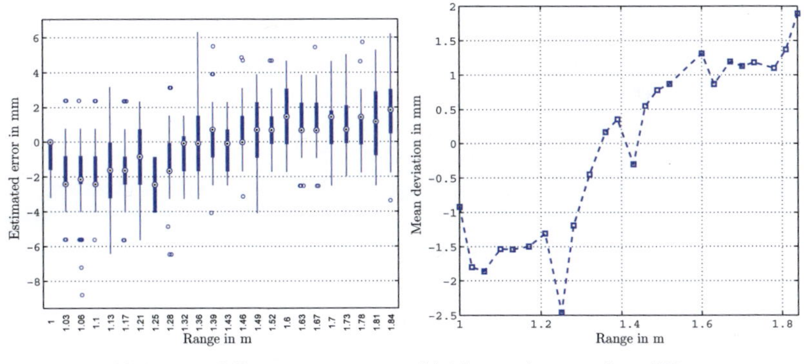

(a) Measured distance

(b) Measured mean value of 20 measurments

Figure 4.15: Distance estimation for single target corner reflector from 1.0 m to 1.84 m.

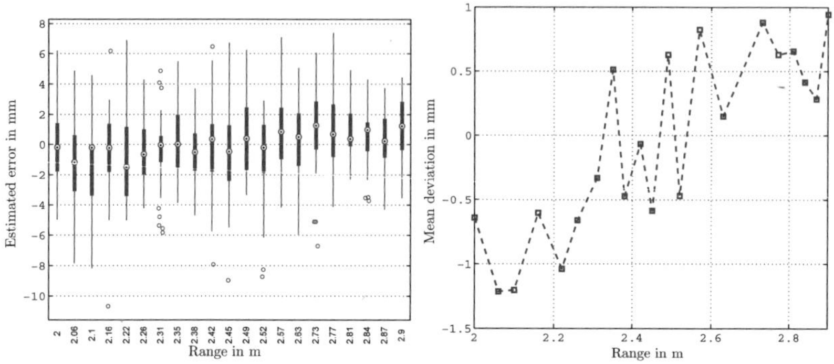

(a) Measured distance (b) Measured mean value of 20 measurments

Figure 4.16: Distance estimation for single target corner reflector from 2.0 m to 2.9 m.

4.6 Heterodyne system with a direct DC conversion

The radar system architecture described in section 4.3 has the flexibility either to make the signal processing at a finite IF, as described in section 4.4, or to down-convert the IF- signals directly to baseband or DC. In order to assess and compare the short range measurement capability of a homodyne and a heterodyne radar architectures, the same radar system[1] is configured to operate in a homodyne mode, which is detailed in this section. In this case, the baseband signal is produced by down-converting the REF and IF signals in analog domain, as shown in Figure 4.17. The down-converted frequency results directly at baseband, thus it is not possible to get ride of the DC-offset issues in the radar systems.

System integration

The radar system comprises PLL-stabilized synthesizers to lock the two on-chip oscillators, as shown in the upper part of Figure 4.17. Commercially available fractional-N-PLLs are used for stabilization and FMCW ramp gen-

[1] Contributions of this section have been partially reported in [GHS+12]. Some of the author's own publication contents are adopted in the text.

eration. The transmit VCO (VCO_{TX}) is stabilized by using a 60 MHz quartz oscillator as a reference. This is done by locking the reference IF output signals, which are already down-mixed in the chip. The VCO_{TX} operates with a small frequency offset to VCO_{RX}, thus the generated reference signal frequency is usually at 600 MHz.

The VCO's divide-by-64 signal, which is on the order of 1.9 GHz, is fed to PLL_{RX} to stabilize VCO_{RX}. As the maximum input frequency of the phase frequency detector (PFD) is 120 MHz, the divider signal is divided-by-32 on the PLL's N-counter. Ramp generation is done by sweeping the reference input frequency of the PLL_{RX}, which is slow enough for the PLL to settle to the desired frequency value. A passive loop-filter is used for both PLL circuits.

The received IF signals are further down-converted to baseband, by multiplying with the reference signals. Since the two oscillators are not correlated, only this off-chip mixing stage will provide the desired phase correlation between the transmit and receive signal. For FMCW modulation, the baseband signal contains information proportional to target distance and Doppler shift. The low-frequency circuits for mixers, baseband amplifiers, differential to single-ended converters and filters are assembled on another two boards, which can be connected to the back side of the RF board. Each channel of the radar IF is first filtered to remove aliases in the processing stage. The filter output drives an amplification stage with selectable gain.

To verify the functionality of the radar module in a realistic scenario, a complete transmission system has been setup and measured. In the case of FMCW radar, the reference oscillator frequency is swept and introduced into the PLL. In this configuration, the transmitter sends a signal, with a FMCW-chirp of 10 ms ramp duration T_m and 820 MHz frequency swing f at a carrier frequency of f_c = 121.6 GHz. During this measurement, the IF signal was set to 600 MHz. The signal processing for the presented measurement took place on a PC. A 12-bit ADC with a dynamic range of maximum 73 dB was used for analog to digital conversion. Hanning weighting function is used prior to the FFT to calculate the target distance from the received spectrum. A peak can be readily detected that corresponds to the reflector. Figure 4.18 show the processed signal obtained from a corner reflector moved on a linear drive in the range of 0.1 m to 0.8 m.

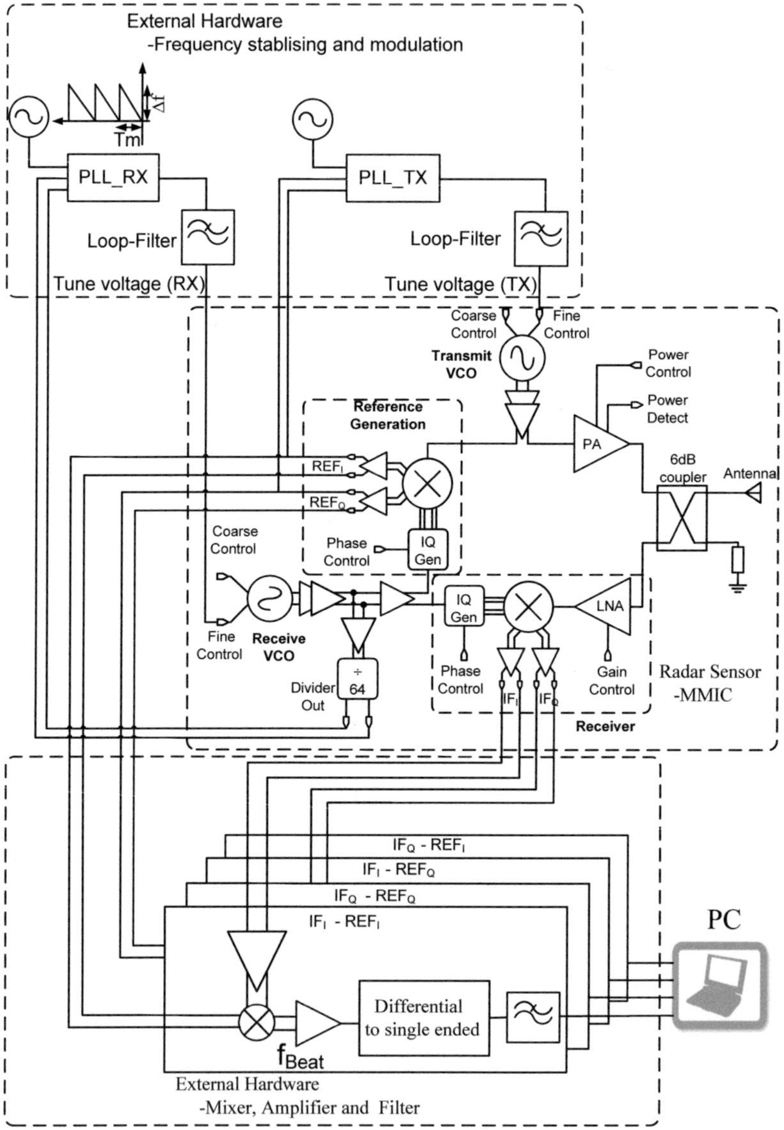

Figure 4.17: Block diagram of experimental radar sensor module. The detailed setup of this system is described in [GHS+12].

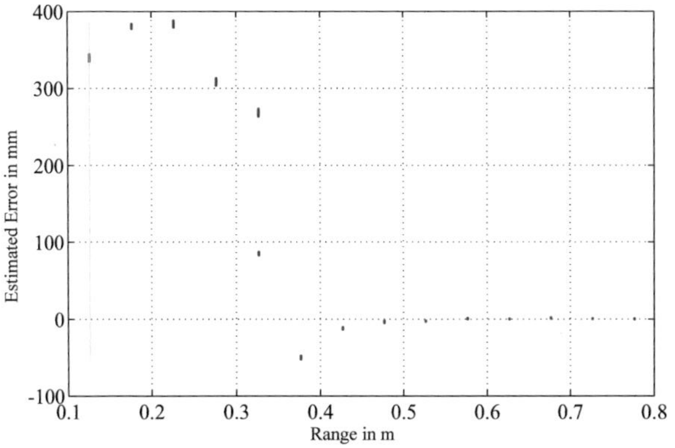

Figure 4.18: Short range distance measurement result of a heterodyne system which is configured as a homodyne. The setup of this system is described in [GHS+12].

4.7 Dynamic DC-offset measurement: case study

In this section, a case study on dynamic DC-offset on radar systems is conducted, based on the two developed systems in chapter 3 and chapter 4. The goal is to analyze and compare the effect of temperature and random noise sources on the short-range measurement performance of the two radar systems. All measurements were done using the set-up and configuration as described in the respective sections. Several measurements with a fixed scenario and no-target are taken in an EMC measurement chamber for a long period of time to analyze the effect of temperature over time. For both systems an empty signal subtraction is applied prior to the FFT. This result demonstrates that the DC-offset in a homodyne system varies over time, where as in a heterodyne system the DC-offset will be eliminated prior to the down-conversion to baseband. Figure 4.19 and 4.20 shows the effect of temperature drift on the DC-offset of a radar system. The DC-offset induced by temperature variation in radar system affects the accuracy and the performance of measurements. This is especially true for short range radar systems in which the beat frequency is close to DC. In the heterodyne system, the dynamic DC-offset which is introduced by a temperature variation could be fully calibrated.

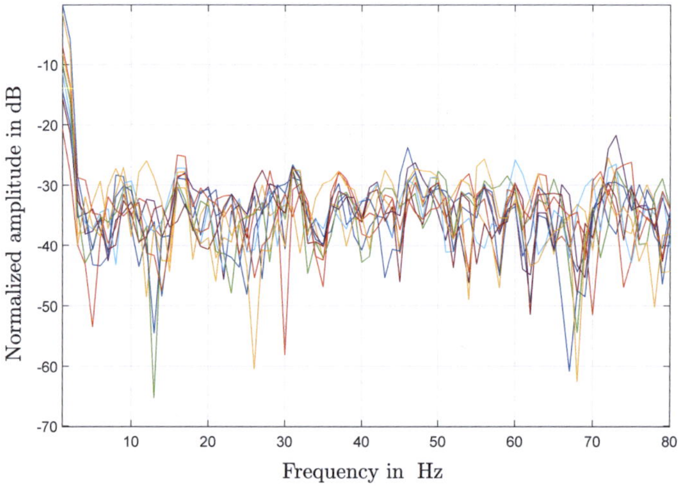

Figure 4.19: Measured DC-offset variation over time after empty signal calibration in homodyne system.

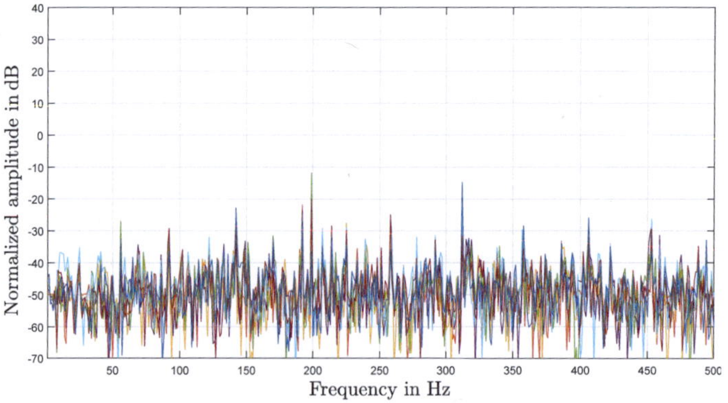

Figure 4.20: Measured DC-offset variation over time after empty signal calibration in a heterodyne radar system.

4.8 Summary and concluding remarks

In this chapter a D-band, single chip, heterodyne radar system for short distance measurement is presented. The developed prototypes have several features superior to the state-of-the-art mm-wave radar prototypes: heterodyne architecture integrated on a single chip with on-chip antenna as well as a flexible system architecture. The presented radar transceiver is based on an IF architecture in order to avoid the specific homodyne problems and enhance the system's sensitivity. As this system has the flexibility to operate in either a heterodyne or homodyne mode, two different sensor configurations are developed, and a comparison of a short range measurement performance of both modes is performed.

In the system described in section 4.4, the signal processing stage is performed at IF frequency. The signal processing at the IF frequency is possible by using the SDR-platform, and the down-conversion to the baseband signal is done in a digital fashion. In that case, there is an advantage to get ride of DC-offset issues at baseband. Where as, in the system described in 4.6, in which the baseband signal is produced by the direct conversion of the IF and REF signals to DC, the DC-offset can not be mitigated.

In general, DC-offset in homodyne radar architectures is more severe than heterodyne architecture, thus the former systems are not appropriate for a short range measurements. However, heterodyne receivers are more complex, since more signal sources and down-conversion stages are required. In spite of these disadvantages heterodyne radar system has shown a better performance for short range measurements. In a case study, it is shown that with a calibration (empty signal subtraction) of a radar signal prior to measurements, it will not be fully possible to get ride of the dynamic DC-offset which usually varies fast over system temperature.

A quantitative short range distance measurement comparison is done between different radar architectures. Results show that a better short range measurement performance can be achieved in the heterodyne configuration compared to the more common homodyne mode due to mainly to the reduced influence of DC-offset mask on the beat signal. Table 4.2 summarizes the comparison of the four radar systems: System-A: A heterodyne system with direct DC conversion, System-B: a heterodyne system with IF- signal processing, System-C:

Range	System-A Section 4.6	System-B Section 4.4	System-C Chapter 3	System-D [GBH+13]
0.1 m - 0.2 m	±400 mm	±2 mm	-	-
0.2 m - 0.3 m	±400 mm	±2 mm	-	-
0.3 m - 0.4 m	±280 mm	±0.5 mm	-	-
0.4 m - 0.5 m	±10 mm	±0.3 mm	-	-
1.0 m	± 4 mm	±1 mm	±2 mm	±2 mm
1.5 m	± 4 mm	±0.5 mm	±2 mm	±2 mm
2.0 m	± 4 mm	±0.5 mm	±2 mm	±2 mm

Table 4.2: Summary and comparison of a homodyne and heterodyne architecture for short range distance measurement: [0.13 - 2 m].

homodyne dual channel monostatic system with on chip antenna, System-D[2]: a homodyne, bistatic radar system, summarizes the comparison of the four radar systems in the distance range of 0.13 m to 2 m, which show a significant performance of a heterodyne radar for short range distance measurements.

[2] This radar system, which is published in [GBH+13], is also the author's own work, which however is not included in the thesis.

5 Embedded auto-calibration technique for radar systems

Systematic errors and imperfections in mm-wave ICs limit the system performance of a radar sensor, especially for short range and precise distance sensing applications. There are numerous sources of systematic errors in a typical integrated radar systems that need to be taken into consideration. These impairments are generated by static reflections of the transmit signal inside the front-end. The reasons are mainly parasitic coupling from the transmit signal to other components due to housing reflection or limited shielding, limited isolation between the transmit and the receive path of the radar coupler, any mismatch between components, the antenna, interconnects and random reflections. All of these parasitic signal components will interfere with the actual received signal, resulting in unwanted signal on the receiver side, which in turn causes a significant degradation in the performance of a radar system.

Besides the static offset and imperfections, also dynamic parasitic components can be observed in the receive signal as shown in Section 4.7. In most cases, they are resulting from the interaction of the environment with the radar sensor system, like temperature effects or aging for very slow drifts, as well as multi-path or multi-target scenarios for the radar channel.

To eliminate the RF-impairments in real-time operation mode, an adaptive concept for online system calibration is necessary. An online radar calibration methodology for short range distance sensing application is implemented in [VW03]. The system application is only for relative distance change measurement which is based on continuous wave (CW) measurement of the reflection coefficient and interpretation of the signal phase. However, there are several application areas where a precise absolute distance measurement is needed. Then, it is necessary to apply a radar calibration procedure to account for any RF-impairments in the system.

A free-space radar calibration method as in [ZSHvdW06] is straightforward, requires no additional hardware and inherently captures all the antenna imperfections, such as its delay and spurious reflections from nearby objects. However, it still requires external components and precise placement of a reflector, which is impossible in many situations. For example, in [SHZ+09] the sensor waveguides and antenna are built in the chassis of a steam turbine, making the use of a reflector difficult. As a result, a built-in calibration method that would minimize or completely avoid the use of external hardware is desirable.

In this chapter, a digitally assisted, embedded-calibration scheme is proposed for mm-wave highly integrated radar systems. The calibration method can adequately compensate non-idealities and systematic errors in radar sensors in analog domain. The solution shows a robust and effective methodology for next generation radar systems where the effects of systematic errors are expected to be much more severe. To the best knowledge of the author, this work shows a complex S-parameter measurement of a load at the antenna port for the first time. The integrated radar sensor described in chapter 6 is used to verify the calibration methodology. The online calibration algorithm and routine is briefly demonstrated.

5.1 Radar architecture with embedded calibration scheme

A single chip D-band, heterodye radar sensor with an auto-calibration scheme and integrated on-chip BIST features is realized. However, the on-chip BIST concept of this radar transceiver will be detailed in the following Chapter 6. The circuit design is realized in a co-operation work with University of Toronto and the chip is fabricated by ST-microelectronic. The mm-wave frontend is based on an integrated circuit realized in a 0.13-μm SiGe-BiCMOS technology with f_T= 230 GHz and f_{max} = 280 GHz [ADC+09]. The circuit blocks of the transceiver have been detailed in [Sar13], so only a brief introduction with an emphasis to the system aspect and the working principle of the sensor is given in this section.

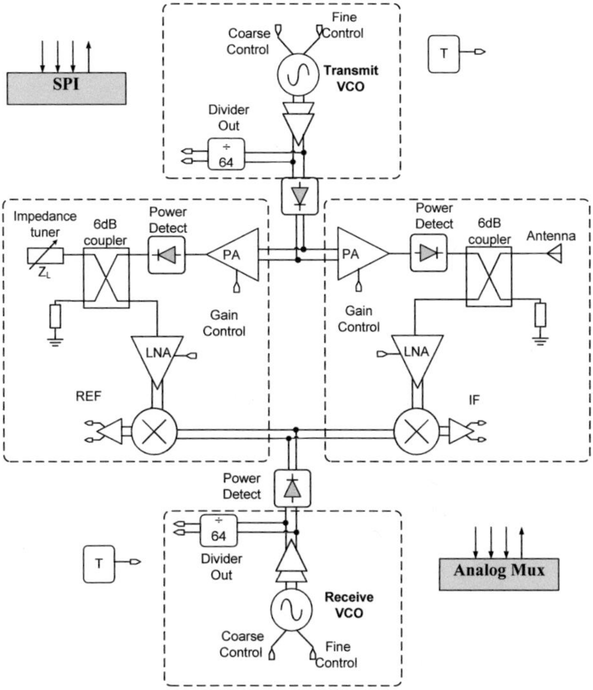

Figure 5.1: System architecture of a 122 GHz radar sensor with integrated BIST features.

5.1.1 System level description

The block diagram of the transceiver chip is shown in Figure 5.1. The transceiver is based on a low-IF frequency with separate transmit and receive VCOs, which oscillate at frequencies f_{TX} and f_{LO}. In the primary channel, the transmit VCO signal is first transmitted by the antenna, reflected by a target whose distance is to be measured, and is received back by the same antenna. The 6-dB coupler separates the transmitted from the reflected signals and steers the reflected signal to the receiver, which in turn, down-converts it to the IF frequency f_{IF} = f_{TX} - f_{LO}. By comparing the IF outputs of the main and of the reference channels, any phase difference (or frequency shift, if frequency modulation is employed) between the transmitted and reflected signals can be accurately resolved to determine the distance to the target. The reference channel is identical to the main, but instead of being connected to an antenna, it is terminated on a variable impedance tuner. The perfect symmetry between the two channels cancels any systematic delay mismatch between them, and thus captures the round-trip delay to the target as accurately as possible. The total size of the chip is 2.6 mm x 2.3 mm, as shown in Figure 5.2. Table 5.1 summarizes the main performance and characteristics of the radar IC.

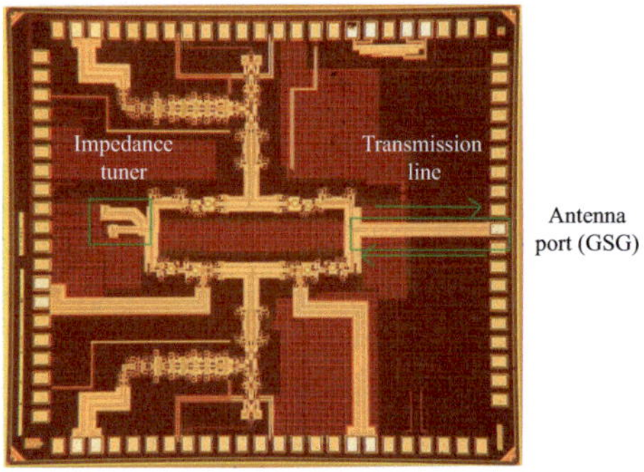

Figure 5.2: Transceiver micrograph with the integrated digital impedance tuner, chip area 2.6 mm x 2.3 mm.

Parameter	**Value/ Range**
Power consumption	952 mW
TX-/RX VCO tuning range	117 - 127 GHz
Maximum transmit power	0 dBm
Antenna gain (simulated)	10 dBi
IF frequency	5 GHz
Receiver gain	13 dB
Noise figure	11.5 dB

Table 5.1: Performance and characterstics of the radar IC.

5.2 Built-in embedded radar calibration concept

An embedded calibration[1] scheme is proposed in this section, which compensates non-idealities and systematic errors in radar sensors in analog domain. In addition, the implemented architecture enables to realize a real time, on-line calibration of the systematic errors. The solution guarantees a simple, low-cost and effective calibration concept for the next generation radar systems where the effects of systematic errors are expected to be much more severe.

The core element for the embedded calibration is the digital impedance tuner, which has been added a the output of the reference channel to calibrate it as a one-port VNA (Vector Network Analyzer). This involves the calculation of the one-port error coefficients, as described below. Therefore, imperfections of the RF front-end (i.e. leakage, phase delay, unwanted reflections of radome, antenna and connectors mismatch etc.) can be estimated and calibrated out. Since the reference and primary channels are identical, the same error coefficients determined from the reference channel can be used to correct the primary channel during an actual measurement. The correction of the measured reflection coefficient of the primary channel will remove the systematic errors in the measured distance to the target introduced by the front-end.

Digital impedance tuner

An impedance tuner was added at the output of the reference channel. This has been fabricated and characterized as a separate breakout and the values of its impedance states are known. The fabricated breakout circuit of the impedance tuner is shown in Figure 5.3. By switching between different known impedance states in a digital fashion, different amplitudes and phase shifts between the reference and IF outputs can be observed, thus verifying the proper operation of the reference channel. This is done by switching the I and Q-indexes of the digital tuner, digitally. Four bits were selected to control the two indexes separately, which results in a total of 256 impedance states of the tuner. Thus, a one port calibration using the known tuner impedance states can be performed and the imperfections of the RF front-end (i.e. leakage, phase delays, etc.) can be estimated and calibrated out. The equivalence between

[1] The embedded calibration concept is patented by the author [14].

the main and reference channel minimizes any delay mismatch between the two, thus capturing the round-trip delay to the target as accurately as possible.

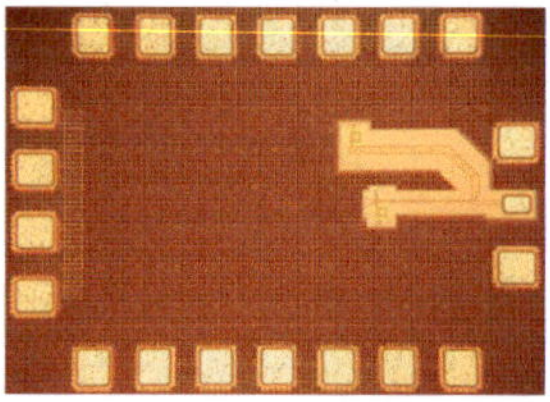

Figure 5.3: Breakout circuit of the impedance tuner.

The complex S-parameter measurement methodology and embedded calibration is detailed in section 5.2. In addition, unwanted reflections due to radome and covered reflections are calculated during the calibration procedure, which can be used to correct the measurement during a normal radar operation. This concept is especially advantageous for automotive radars at 77 GHz, which are equipped with a radome and are usually integrated behind a bumper.

5.2.1 Error box model of RF-impairments in radar system

Figure 5.4 shows a simplified schematic representation of RF-impairments presented in the radar front-end. A summary of all describing parameters is listed in Table 5.2. In a VNA system, a calibration process characterizes most of the VNA systematic error. Therefore, before performing the error correction via VNA calibration, it is important to model the systematic error, which is knows as error box model.

Similar to a one port VNA, all RF-impairments can be summarized and modeled as in the signal flow diagram shown in Figure 5.5. The systematic errors can be mathematically represented as in Equation 5.1, in which the error terms are defined as: e_{00} as measurement system directivity error, e_{11} as measurement system source match error, and the terms $e_{01}e_{10}$ is the reflection tracking

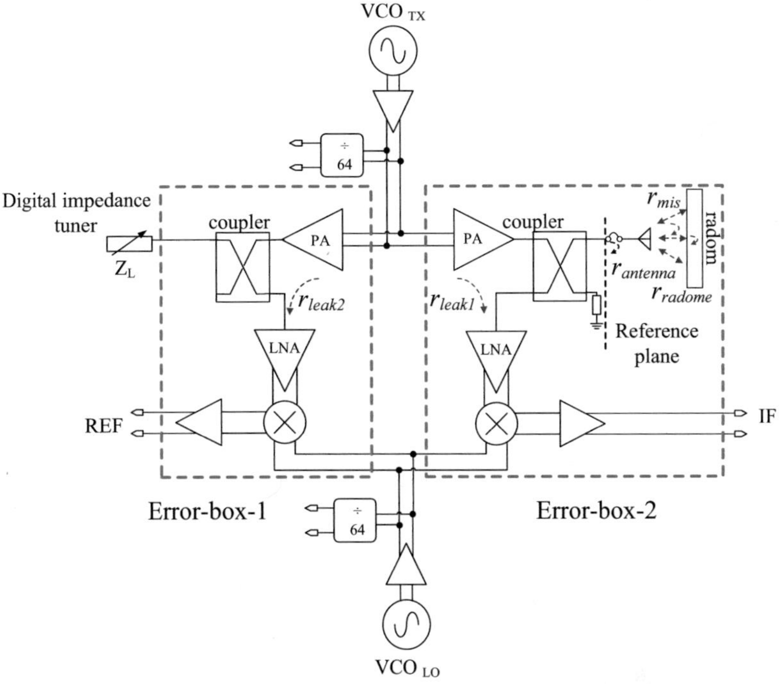

Figure 5.4: Simplified schematic of systematic error sources in the radar front-end.

Symbol	Quantity
r_{leak1}	RF-LO leakage in the main channel,due to imperfect coupler isolation
r_{leak2}	RF-LO leakage in the reference channel, due to imperfect coupler isolation
$r_{antenna}$	Reflection due to antenna and connector mismatch
r_{mis}	unwanted multiple reflection
r_{radome}	direct reflection on the surface of radome

Table 5.2: Summary of symbols used to describe RF- impairments in the radar front end.

error. Solving the one-port flow graph yields a bilinear relationship between the actual, Γ_L, and measured, Γ_M reflection coefficient.

$$\Gamma_{Error} = \begin{bmatrix} e_{00} & 1 \\ e_{10}e_{01} & e_{11} \end{bmatrix} \tag{5.1}$$

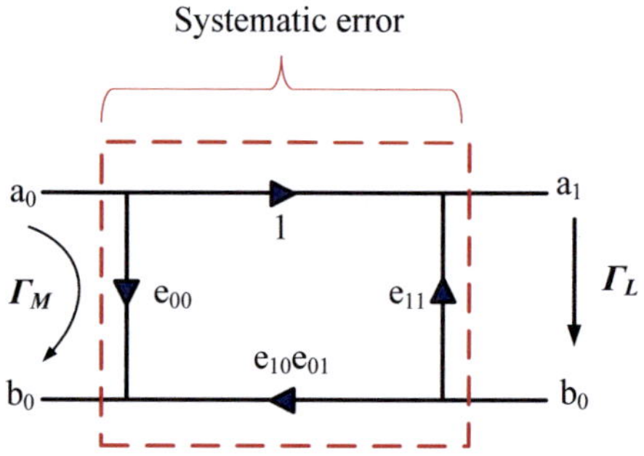

Figure 5.5: One-port error-box model signal flow diagram representation of systematic errors.

5.2.2 Calibration methodology

The procedure of the systematic error extraction and calibration process is summarized in Figure 5.7. The impedance tuner can be controlled using I and Q indexes digitally, which results independent control of magnitude and phase of the reflection coefficients, as shown in Figure 5.6. The output of the reference and main channels are used to calculate the complex load impedance. A real part and an imaginary part of a load impedance is calculated based on voltages measured at the output of IF and REF channels and on known values of the tunable digital impedance circuit. Assume the voltage output of the reference channel and main channel can be represented by the following equations.

$$V_{REF} = A_{REF} \cdot e^{(2\pi f t \varphi_1)} \tag{5.2}$$

$$V_{IF} = A_{IF} \cdot e^{(2\pi f t \varphi_2)} \tag{5.3}$$

For any two measured REF, V_{REF} and IF , V_{IF} output voltages, the reflection parameter Γ_M can be computed as in Equation 5.4.

$$\Gamma_M = \frac{V_{IF}}{V_{REF}} = \Delta A \cdot e^{j\Delta\varphi} \tag{5.4}$$

The digital impedance tuner provides 256 different points distributed on a

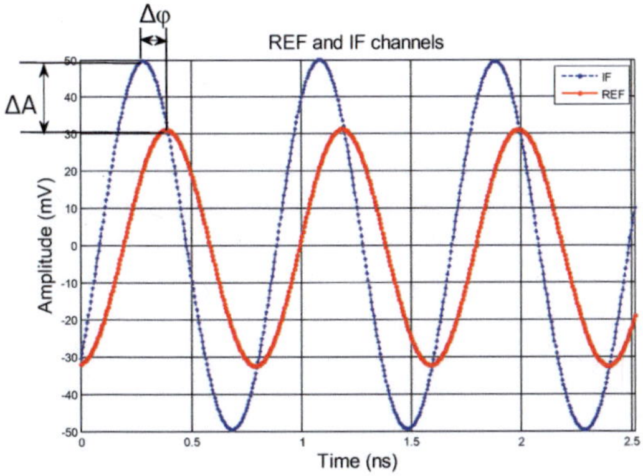

Figure 5.6: Phase and amplitude difference measurement from measured IF and REF signals.

smith chart, which are all frequency dependent. To include and calculate all error parameters, at least 3 measurements are necessary, which can be obtained by changing the complex digital impedance tuner states ($\Gamma_{L1}, \Gamma_{L2}, \Gamma_{L3}$) digitally. The corresponding reflection coefficients, Γ_{M1}, Γ_{M2}, and Γ_{M3}, can be calculated using equation 5.4. By solving the resulting linear equation, the error coefficients can be computed as follows:

$$\begin{bmatrix} 1 & \Gamma_{L1}\Gamma_{M1} & -\Gamma_{L1} \\ 1 & \Gamma_{L2}\Gamma_{M2} & -\Gamma_{L2} \\ 1 & \Gamma_{L3}\Gamma_{M3} & -\Gamma_{L3} \end{bmatrix} \cdot \begin{bmatrix} e_{00} \\ e_{11} \\ e_{\Delta} \end{bmatrix} = \begin{bmatrix} \Gamma_{M1} \\ \Gamma_{M2} \\ \Gamma_{M3} \end{bmatrix} \tag{5.5}$$

where

$$e_{\Delta} = e_{00}e_{11} - e_{10}e_{01} \tag{5.6}$$

After calibration, any load Γ_L connected either to the reference or to the main channel can be calculated from the measured Γ_M using:

$$\Gamma_L = \frac{\Gamma_M - e_{11}}{e_{00} - e_{\Delta}\Gamma_M} \tag{5.7}$$

If no target is present ($\Gamma_M = 1$), Γ_L becomes,

$$\Gamma_{Error} = \frac{1 - e_{00}}{e_{11} - e_{\Delta}} \tag{5.8}$$

For regular measurements, i.e. when a target is present, the tunable load is set to the calculated Γ_{Error} and the reference channel becomes a calibration for the main channel. To set the value of the impedance tuner to the calculated Γ_{Error}, a look-up table can be used. After calibration, while the measurement doesn't change, the phase and amplitude difference of the two channels is zero. Thus the systematic errors due to all the impairments is compensated. This procedure may be repeated as required, if there is any environmental change, such as: temperature, integration, adjustment etc. In this case, it is assumed that the complete radar transceiver is integrated in one chip, so there is no process variation between the reference channel and the main channel, and they are absolutely symmetrical.

Although, three measurements are sufficient to calculate the unknown error coefficients, increasing the number of measurements leads to an accurate and robust calibration parameters. In this case, the number of measurement points is chosen to be larger than the number of the unknown coefficients a, b and c to obtain a good fit. Since measurement redundancy is advantageous for verifying the technique, the chip has been calibrated in all its 256 states available and the unknown error coefficients a, b and c are obtained. Then the matrix

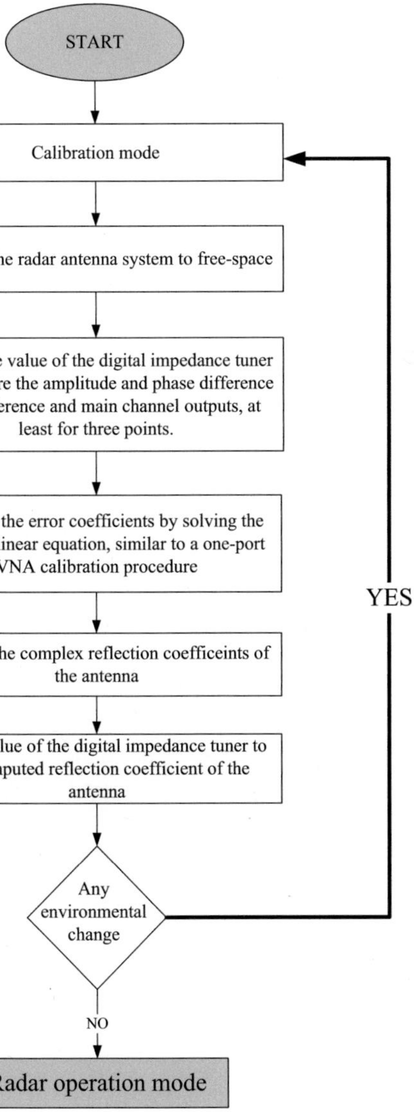

Figure 5.7: Steps of error coefficient extraction and calibration

equation is overdetermined and only approximate solutions can be found. An often used criterion is to minimize the square of the error:

$$\begin{bmatrix} \Gamma_{L1} & 1 & -\Gamma_{L1}\Gamma_{M1} \\ \Gamma_{L2} & 1 & -\Gamma_{L2}\Gamma_{M2} \\ \Gamma_{L3} & 1 & -\Gamma_{L3}\Gamma_{M3} \\ \vdots & \vdots & \vdots \\ \Gamma_{Ln} & 1 & -\Gamma_{Ln}\Gamma_{Mn} \end{bmatrix} \cdot \begin{bmatrix} a \\ b \\ c \end{bmatrix} = \begin{bmatrix} \Gamma_{M1} \\ \Gamma_{M2} \\ \Gamma_{M3} \\ \vdots \\ \Gamma_{Mn} \end{bmatrix} \tag{5.9}$$

After computing the error parameters: a, b and c, the actual reflection parameter or complex load can be computed from each measured parameter, by using equation 5.10.

$$\Gamma_L = \frac{\Gamma_M - b}{a - c\Gamma_M} \tag{5.10}$$

This embedded radar calibration method thus achieves a significant time saving and is well suited for iterative on-line monitoring and calibration of radar systems. Note that the values of the impedance tuner are frequency dependent.

5.3 System integration

Consider the simplified block diagram depicted in Figure 5.8, which is based on a low-IF frequency heterodyne architecture achieved by using separate transmit and receive VCOs. Two separate PLL circuits are necessary to lock and stabilize the two VCOs, as well as to modulate the transmit signal in the case of FMCW mode. The system is built-up by designing a dedicated test-board for the power supply and digital control of the integrated MMIC. A QFN packaged chip (without antenna[2]) is mounted on the test-board, which allows an on-wafer probe measurements. Since this system is implemented for the verification of embedded- radar calibration methodology, the antenna port is left open to flexibly probe a know impedance values for concept verification. The test-board contains, a power supply, an FTDI- Module for the on-chip SPI programming, and several coaxial connectors, to capture the IF signals coming out of the chip, and which will be displayed on oscilloscope. The IF and

[2] The same radar system with integrated mm-wave antenna is detailed in Chapter 6.

REF signals are sampled by a digital oscilloscope from the oscilloscope and are acquired on a computer for further processing.

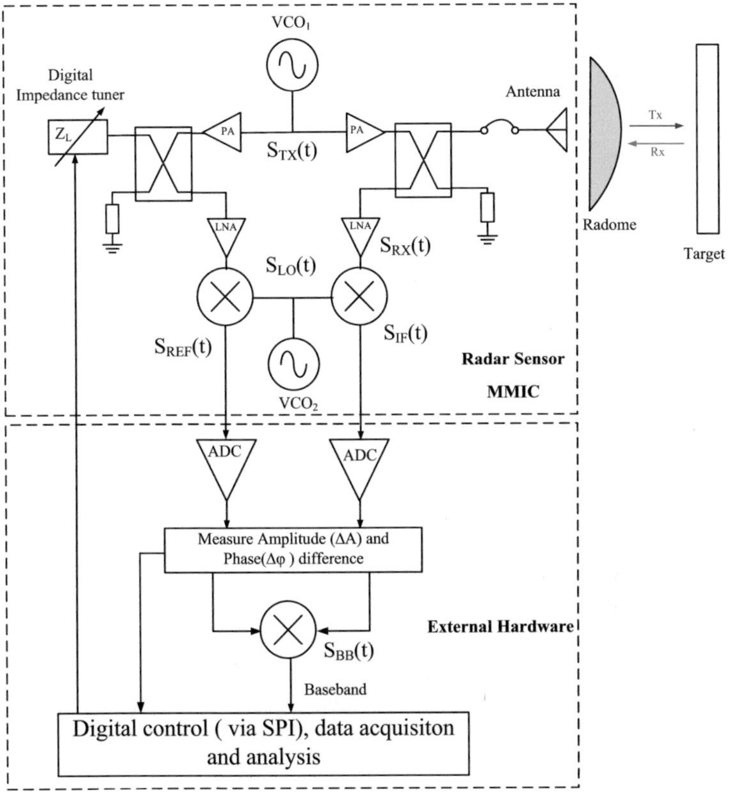

Figure 5.8: A simplified system diagram illustrating the embedded calibration concept based on the the radar system architecture shown in Figure 5.1.

Signal path flow

Assume the oscillation frequencies of the transmit VCO and the local oscillator or reference VCO, are f_{TX} and f_{LO} respectively. The signal generated by each oscillator is distributed to both the reference and the main channel using active power splitting. In both transceiver channels, the receive VCO signal drives the down-convert mixers while the transmit VCO signal drives the power amplifiers with a maximum output power of 0 dBm.

The transmit signal at the output of the power amplifier can be expressed as:

$$S_{TX}(t) = A_{TX}cos(2\pi\, f_{TX}(t)) \tag{5.11}$$

The received signal at the input of the receiver mixer after the round-trip delay τ_D is:

$$S_{RX}(t) = A_{RX}cos(2\pi\, f_{TX}(t - \tau_D)) \tag{5.12}$$

which is given by $\tau_D = 2R/c$, where R is the range to the target, and c is the speed of light. And the IF signals at the output of the reference and main receiver are:

$$S_{REF}(t) = A_{REF}cos(2\pi(f_{LO} - f_{TX})(t)) \tag{5.13}$$

$$S_{IF}(t) = A_{IF}cos(2\pi(f_{LO} - f_{TX}(t - \tau_D)) \tag{5.14}$$

To obtain the baseband signal or the beat frequency a further down-conversion stage is required.

$$S_{BB}(t) = A_{BB}cos(2\pi\, f_{TX}\tau_D) \tag{5.15}$$

The reference channel is designed to be identical to the main channel that performs the actual measurement, but instead of being connected to an antenna, it is terminated on a variable impedance tuner. By comparing the IF outputs of the main and reference channels, any phase or frequency shifts between the transmitted and reflected signals can be resolved. The baseband signal is obtained by down converting (multiplying in digital domain) the S_{IF} and S_{REF} signals. Most importantly, since the phase noise of the two VCOs appears at both outputs, they will be correlated during this mixing process, which significantly increases the system accuracy by minimizing impact of phase noise. Figure 5.9(a) and 5.9(b), shows a typical time signal out put of the reference and IF channel, which is produced by tuning the I and Q indexes of the impedance tuner. The antenna port is left open, and the measurement environment doesn´t change to keep the IF output constant as seen in Figure 5.9(b). For the calibration process, the digital impedance tuner is tuned with different state values, which in-turn change the REF output signal, as displayed in Figure 5.9(a). By comparing the IF and REF signals, the resulting amplitude and

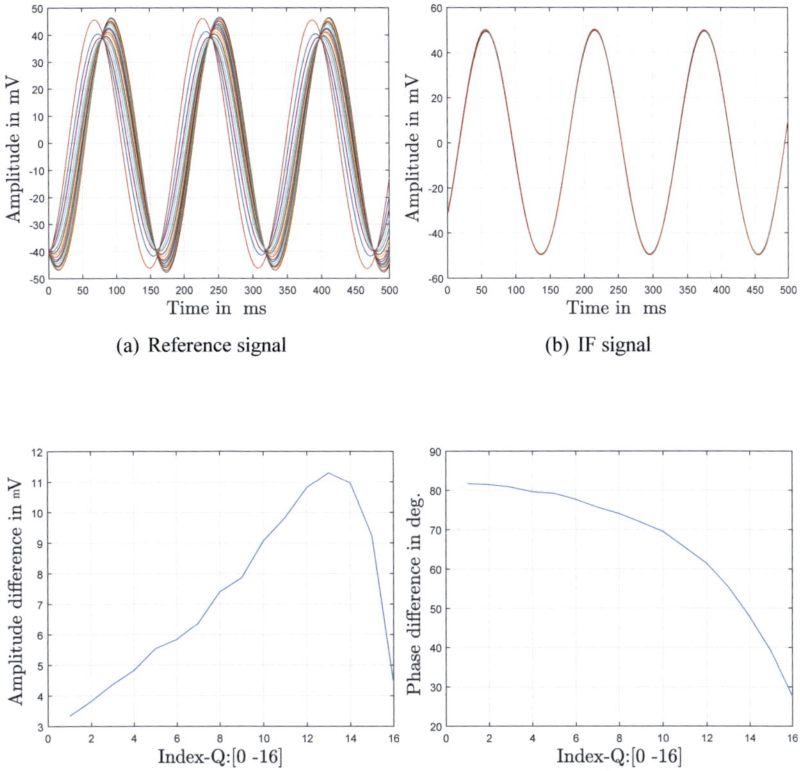

(a) Reference signal

(b) IF signal

(c) Amplitude difference

(d) Phase difference

Figure 5.9: Measured time signal of REF and IF signal in CW mode, for a calibration mode.

phase variation is displayed in Figure 5.9(c) and 5.9(d) respectively.

For all combination of I and Q- indexes, which is 256 measurements, a voltage amplitude variation range of 20 mV to 55 mV and a phase variation ranging from -90° to -5° is achieved. A sample plot of magnitude and phase variation at 122 GHz frequency is plotted in Figure 5.10 and 5.11, respectively. During the measurements, the antenna port is left open circuited and the measurement environment didn´t change.

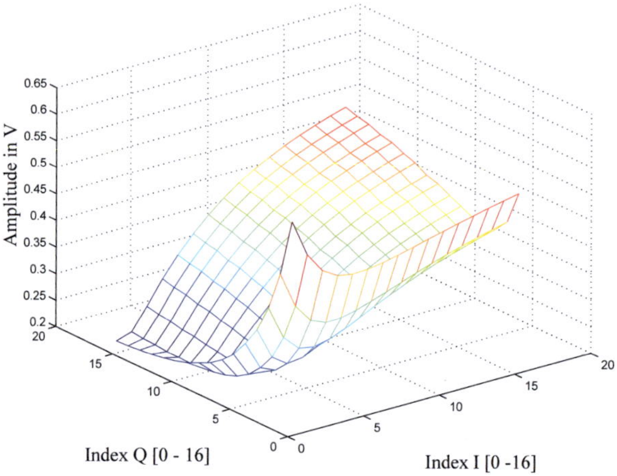

Figure 5.10: Magnitude variation of S-parameter at 122 GHz by tuning the I and Q values of the complex impedance tuner.

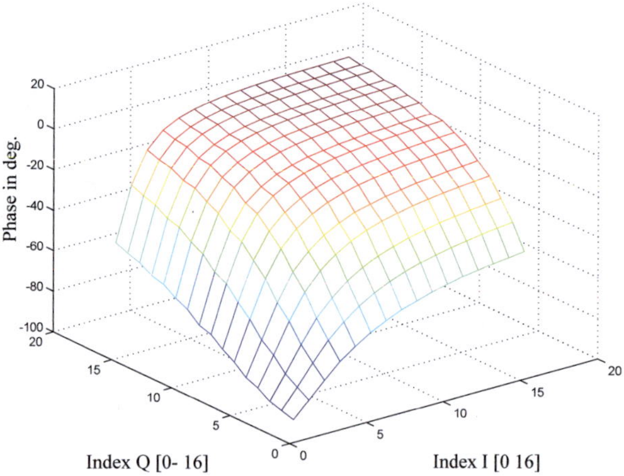

Figure 5.11: Phase variation at 122 GHz by tuning the I and Q values of the complex impedance tuner.

5.4 Validation of the proposed methodology

To validate the proposed embedded calibration methodology, mm-wave measurements were performed by using a variable complex impedance tuner. The measurement procedure outlined in this section uses Vector Network Analyzer (VNA) measurement techniques for on-wafer S-parameter measurements. To determine the S-parameter characteristics of the DUT (Device Under Test), a one-port S-parameter measurement is performed in lab, as shown in Figrue 5.12(a). In the first step, characterization measurements of the variable complex impedance tuner were conducted between 90 GHz and 140 GHz by using a HP8510C network analyzer interfaced with OML T/R module and GSG wafer probes from Cascade Microtech. A variable phase shifter combined with a variable attenuator which is terminated by a short was used to produce different complex load impedance which varies over frequency, as shown in Figures 5.12(a) and 5.12(b). This is done by calibrating the VNA to the wafer probe tips using a GGB CS-15 calibration substrate in the intended frequency range. By using a THRU-line on the calibration substrate, the S-parameter of the DUT is measured at different screw positions. The S-parameters of the DUT is determined by de-embedding the THRU-line effect from the measurements, which is modeled as a loss-less delay line. The THRU line model for de-embedding, assuming that it is loss-less is:

$$S^{THRU} = \begin{bmatrix} S_{11} & S_{12} \\ S_{21} & S_{22} \end{bmatrix} = \begin{bmatrix} 0+j0 & e^{-j2\pi f.delay} \\ e^{-j2\pi f.delay} & 0+j0 \end{bmatrix} \tag{5.16}$$

where f is frequency and the delay = 1.13 ps calculated for a THRU line length of 175μm. Thus, the S-parameter of the DUT can be determined by:

$$S^{DUT} = \frac{S_{11} - S_{Measured}}{S_{11}S_{22} - S_{22}S_{Measured} - S_{12}S_{21}} \tag{5.17}$$

Figure 5.13 and 5.14 shows the characterization measurement of the variable attenuator and the phase shifter for the frequency range of 110 - 130 GHz, respectively. The complex impedance of the two devices combined with on-wafer probe tips, is characterized over frequency using a vector network analyzer for different screw positions.

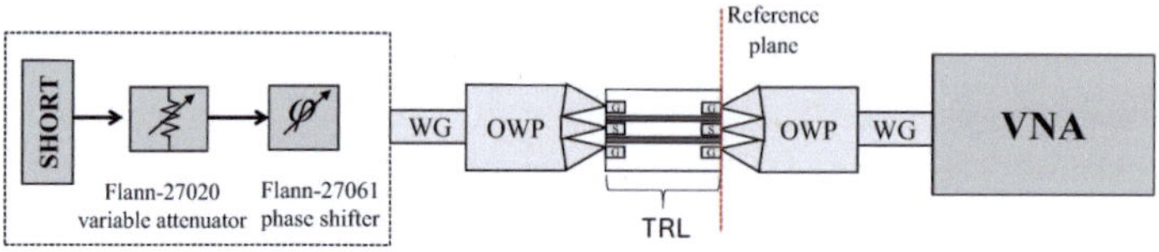

(a) Simplified representation of the measurement set-up.

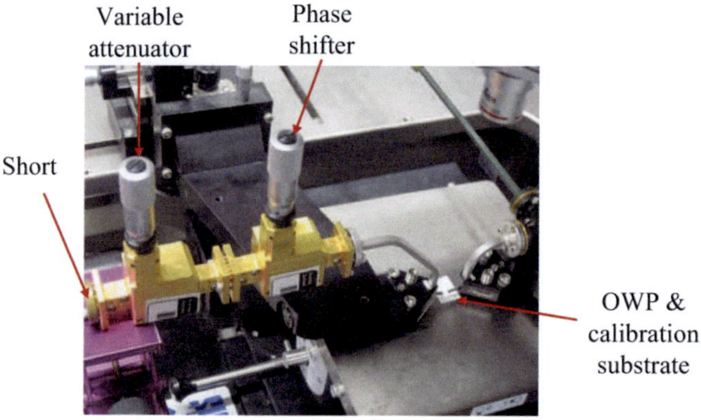

(b) A picture of the one-port VNA measurement.

Figure 5.12: Measurement set-up and photo of the one port VNA to characterize the variable complex impedance.

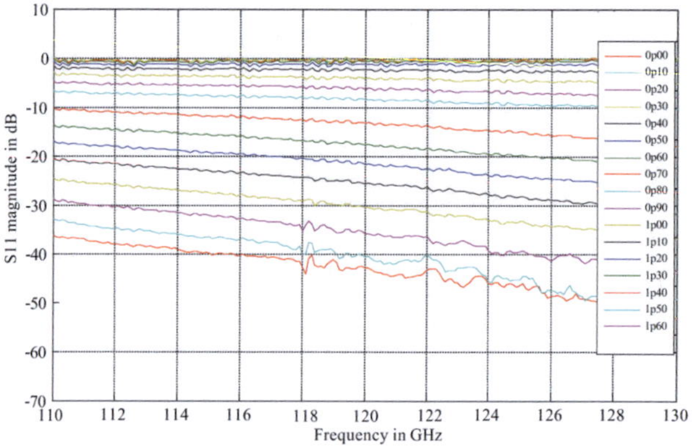

Figure 5.13: Amplitude variation of the variable attenuator at various screw positions.

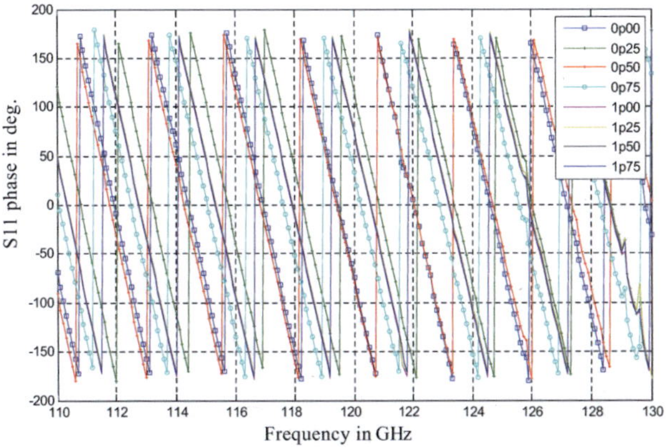

Figure 5.14: Phase variation measurement of the phase shifter at various screw positions.

On-chip measurements

The on-chip measurement of the complex S-parameter is performed by probing the characterized DUT at the radar MMIC antenna port, as shown in Figure 5.15 and 5.16. Prior to this, the values of the complex load impedance, S^{DUT}, are characterized over frequency for different screwing positions as described in the section 5.4. The frequency difference between the two VCOs is set to 1.25 GHz, which is chosen to avoid injection locking between the two VCOs. The divider outputs are used to adjust and tune the transmit VCO to the frequency of interest. The IF and REF signal voltage outputs are directly captured, to compute the reflection coefficients as described in Equation 5.4, which is then used to compare with the reflection coefficients which are acquired by the VNA measurements. However, since the two way measurements of the on-chip transmission line and the capacitive effects of the Ground-Signal-Ground pads is not considered, their S-parameters should be de-embedded to extract the reflection coefficient of the sliding-short. S-parameter simulation of the on-chip transmission line model, which is 750μm long, and the antenna pads is performed in ADS. The simulation values are de-embedded to get the net S-parameter of the DUT.

Using the methods described in Section 5.2, the available complex impedance from the stand alone tuner, the raw on-chip measurements before and after correction are mapped on smith chart as shown in Figure 5.17. Figure 5.18,

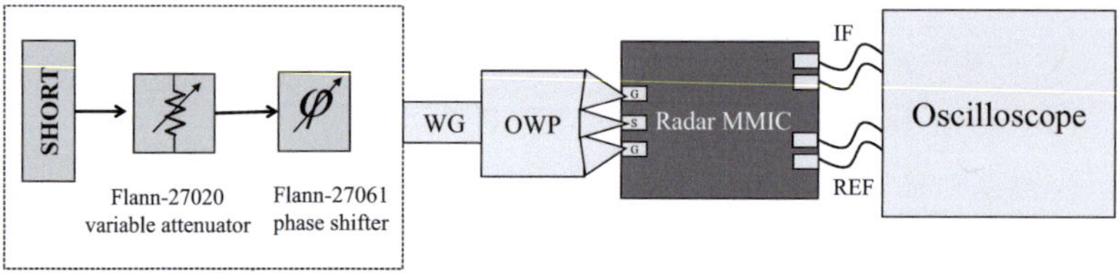

Figure 5.15: Simplified representation of the on-chip complex S-parameters measurement set-up.

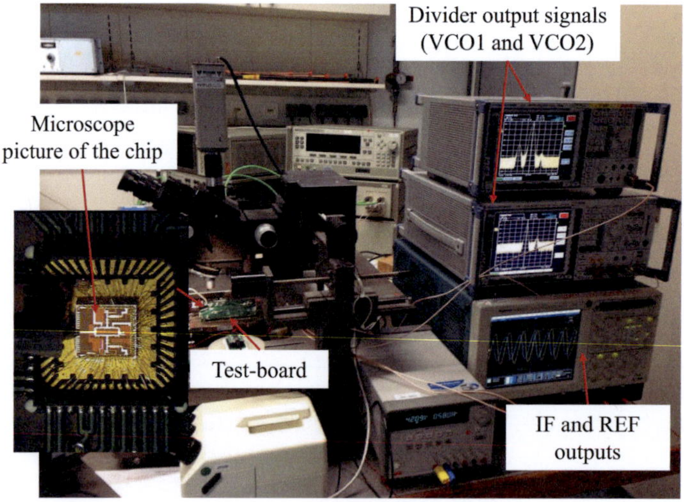

Figure 5.16: Experimental set-up of the OWP measurements and the microscope picture of the mounted chip. The IF and REF output signals are sampled by using a digital oscilloscope, for further processing. Two spectrum analyzers are used to monitor the VCOs signal which can be tuned and adjusted as necessary.

shows the amplitude deviation of the estimated chip load-impedance and the VNA values, which shows a magnitude error range of $\pm$ 0.5 dB.

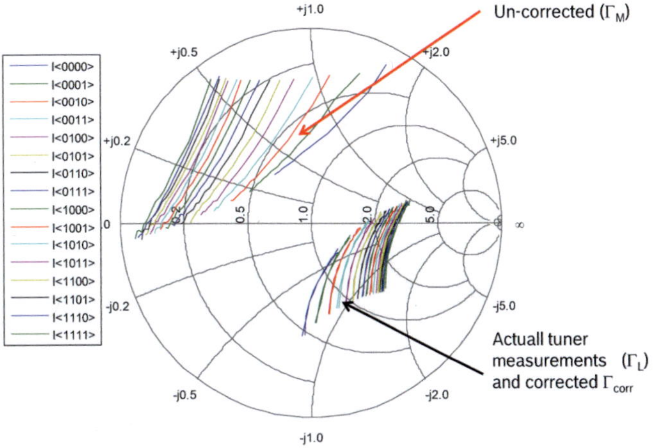

Figure 5.17: Uncorrected chip measurements (Γ_M), impedance states provided by the on-chip tuner (Γ_L), and corrected reflection coefficients Γ_{corr}) at 122.5 GHz

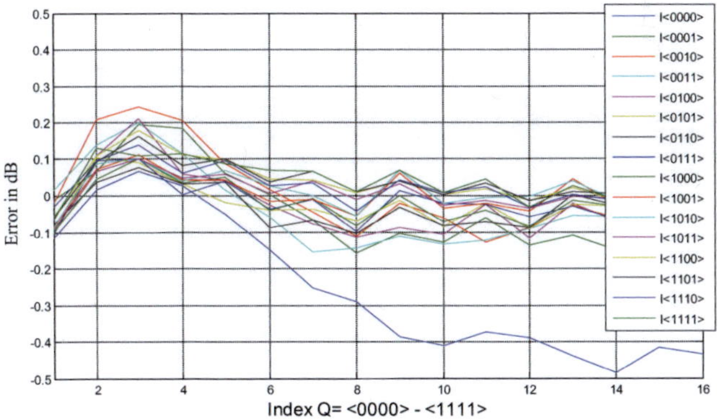

Figure 5.18: Magnitude of amplitude deviation between chip and VNA measurements at 122.5 GHz

The experimental results shown in this measurements show a good agreement of VNA and on-chip measured results. Figure 5.19 and 5.20 shows, the amplitude and phase comparison of the two measurements for different screw position of the variable impedance tuner. The complex impedance measurement is plotted in Figure 5.21. Moreover, it was found that the phase component of these errors dominate compared to the cases of the magnitude of the reflection coefficients.

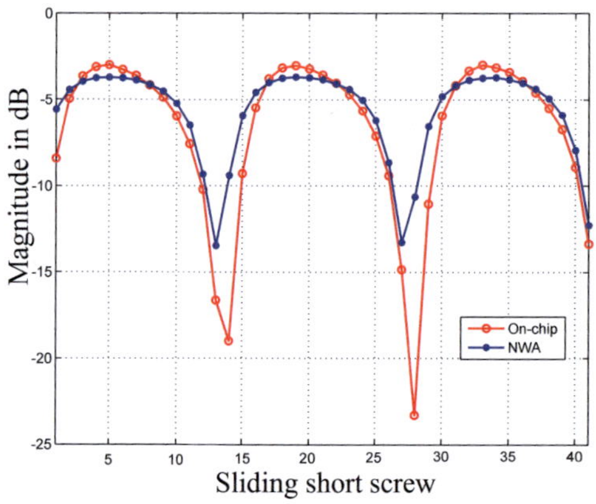

Figure 5.19: Magnitude comparison of network analyzer and on-chip measurements

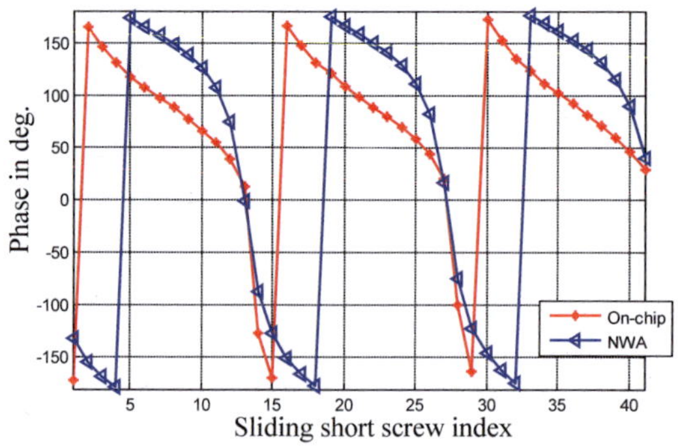

Figure 5.20: Phase comparison of network analyzer and on-chip measurements

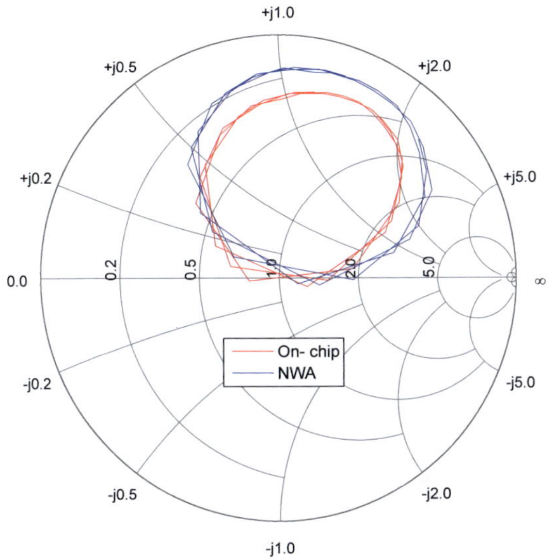

Figure 5.21: Complex impedance comparison of NWA and on-chip measurements.

5.5 Concluding remarks

A novel, real-time, embedded calibration scheme of a radar transceiver chip with an integrated reference channel is presented in this chapter. A digital impedance tuner is used to generate a known complex impedance values, which are later used to estimate the load at the antenna port. A one port VNA methodology with 256 measurements is employed to characterize the error coefficients. Imperfections such as offset, LO leakage, and/or delay mismatch across the system are thus determined, thus can be calibrated out using the computed error coefficients. The main benefit of this methodology is, the calibration is implemented in analog domain prior to the FFT processing, thus it highly increases the SNR of the system, compared to the state-of-the-art methodologies. The proposed calibration method has been verified by measurements, which shows a very good comparison between VNA and on-chip measurements. This methodology, can be further developed for an on-line error calibration of future radar sensor systems, especially for applications which require an accurate and precise measurements.

6 A D-band radar sensor with Built-in-Self-Test

Advances in mm-wave technologies have enabled the integration of complete transceivers in a single chip, which provides a significant reduction in manufacturing cost. However, the cost of testing and diagnosis of these devices remains the same. Their diverse specifications and high operating frequency make necessary extensive tests, both complex and expensive to perform.

For the testing of ICs, one of the main problems with a high level of integration is the lack of access to each individual building block in the chip. This is because fewer signals are being routed out of the chip, and thus it is not possible to access all of the signals from the integrated components. Therefore, special testing features and nodes need to be implemented for testing of the ICs. Furthermore, when the IC is once integrated into a SoC, an additional source of errors will be introduced because of the integration issues, such as substrate coupling, mismatches, cross-talk, leakage etc.

Integration of on-chip BIST features in mm-wave frequencies will not only reduce the testing cost, but also allow the radar sensor system to be monitored online after integration and assembly. Therefore, it is feasible to reason about integrating testing features and nodes already in early phases of mm-wave IC designs.

This chapter deals with an integrated mm-wave testing concept and methodology for a mm-wave radar sensor system. A D-band radar sensor with on-chip integrated BIST features is proposed and implemented. First, testing sequences and methodology is described briefly. Then, a concept to measure the matching condition of an assembled antenna by measuring the complex S-parameter is shown for the first time.

6.1 Mm-wave IC test challenges

Mm-wave testing is challenging because the analog and RF parameters are difficult to measure due to signal sensitivities and non-deterministic nature at high frequencies. Analog and RF circuits are sensitive to the addition of test circuitry which might load the nodes to be measured and affect the overall performance, so each circuit requires an individual test strategy. In spite of the considerable studies in the past, the suitable test techniques for RF-circuits still remain to be the major bottleneck for mm-wave systems, due to the following challenges.

Cost: Testing of mm-wave integrated transceivers is a complex, time consuming and costly task. RF circuits require the use of complicated test procedures, lengthening test times and increasing the cost of automatic test equipments (ATE). Furthermore, the increase of operational frequencies leads to the fact that the testing of the MMICs can account for up to 70% of the total manufacturing costs and time, and is impacted significantly by the cost of testing the various embedded analog and RF circuit components, [DKW10]. In order to avoid these complex test procedures the integration of an embedded test is highly desirable.

Increasing complexity: Test complexity will increase highly with the level of integration in SoC. Increased circuit complexity on a single chip prevents testing access to internal nodes. The alternative of routing the internal nodes to external pins significantly increases the overall manufacturing cost. Along with the integration, technology scaling has been accompanied by increased device performance and sensitivity to process variations.

DUT specification: In recent years, the specification of mm-wave circuits increased in terms of band-width, operating frequency and performance etc. Each specification test requires a different test configuration, resulting in long test time on expensive ATE.

Calibration and de-embedding: RF test setups require complicated calibration processes. During testing, the main concern is the uniformity of off-chip interconnects and the interface with the ATE, probes, sockets, and cables connected to the measurement systems. This has direct influence on measured behavior, and an accurate measurement can be obtained by a de-embedding procedure. The goal of de-embedding is to find the losses between the DUT and the ATE system and calibrate them out from the measurements. The losses are used to offset the value of an RF measurement obtained by the ATE system.

6.2 State-of-the-art: mm-wave testing

Available RF testing approaches fall into three main categories: on-Wafer test, RF Automatic Test Equipment (ATE) and integrated test techniques. Current practice is to combine all three approaches because no single solution can meet the desired quality, cost, time, and reliability requirements.

On-wafer-test

On-wafer test is a fundamental step performed on each chip present on the wafer, consisting in testing each chip for defects. The standard test equipment is used like probing stations, RF and baseband signal generators, and vector signal analyzers. This approach is perfect for the prototype testing but not suitable for integrated transceiver production testing. Major reasons are the number of extra test pins, cost of the test equipment, and long setup time for test, [Lau02]. A wafer probe must perform a number of difficult mechanical tasks while maintaining excellent high frequency electrical performance. Figure 6.1(a) shows a typical wafer level testing of an RF-IC. The increasing complexity of RF-ICs is creating an obstacle in the implementation of a wafer level test. This is due to all the signal nodes are not accessible for measurements. In addition, high frequency signals cannot be driven off the chip without degradation.

Automatic Test Equipment

In conventional RF Integrated Circuit testing, ATE is used to measure the performance characteristics of the DUT. Upon conclusion of such tests, the DUT is classified as acceptable if each characteristic lies within an acceptable range, according to specifications. Although these measurements are simple, they require a variety of test resources which, together with the long test application times, increase the total manufacturing cost. The ATE should have all the required bandwidth, the signal generation, and analysis capabilities needed for the particular RFIC standard under test. A typical ATE consists of a worksta-

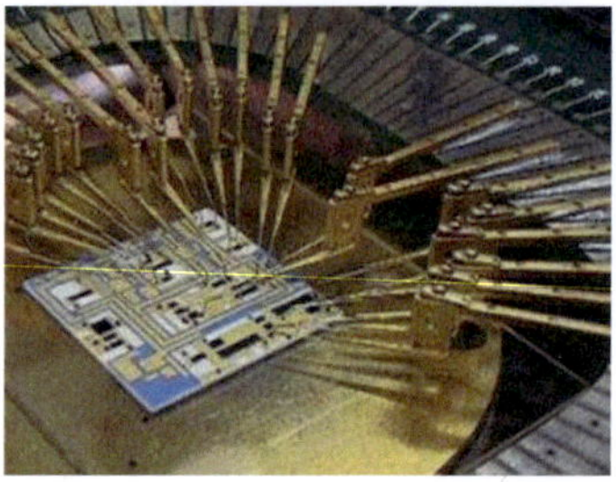

(a) On-wafer IC test

(b) RF-probe card

Figure 6.1: RF testing approachs (cascade microtech).

tion, a test head and several measurement instrumentation. RF-probe cards, as shown in Figure 6.1(b), provides the electrical contact between the DUT and the measurement instruments in the ATE. The types of basic RF stimulus and measurement capabilities include S-parameters, noise figure, power and spectrum analysis. A key problem is that it is not always possible for the ATE to have a direct access to all or even part of the internal signals of an IC, especially in System-on-Chip (SoC) or System-in-Package (SiP) designs. The main limitation and challenge for ATE tests are mainly the high frequency contacts and probes at mm-wave.

Integrated test solutions

A classical approach to reduce the equipment and time cost required for the test procedure, is to embed part of the test resources into the circuit itself. This approach is known as Design for Testability (DfT) or Built-In-Self-Test in case of self-testing. The principle of integrated test is shown in Figure 6.2. On architecture side, a circuit with DfT features incorporate ADC, DAC, and detectors integrated with the analog RF-components. These techniques either directly measure the circuit performance on-chip or produce a signature (e.g. a voltage/ current) that has strong correlation to the circuit. The DfT techniques have a potential to solve or relax the RF-testing problem for future transceivers.

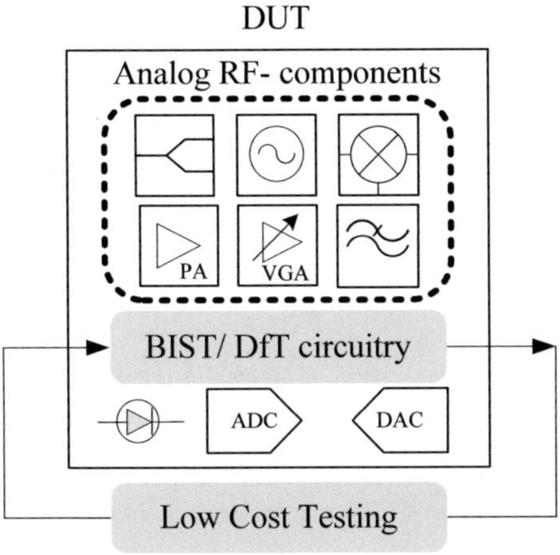

Figure 6.2: Simplified integrated test concepts.

Loopback methods

Loopback testing is a system level test approach for transceiver circuits, that implement both the transmit and the receiver chain onto a single chip, as described in [OSMFSS09]. As shown in Figure 6.3, it involves generation of a test signal in the DSP, digital-to-analog conversion, and up-conversion to the RF frequency in the transmitter chain. Conventionally, the output of the transmitter is then routed back to the input of the receiver, where it is amplified, down converted to the baseband frequency. This signal is converted back to the digital domain and analyzed for functionality verification of the complete transceiver. This method does not require external stimuli and is effective to detect catastrophic faults in the complete signal path. Loopback is one of the low-cost test methods for verifying functionality of a transceiver circuit. However, this testing methodology gives a little insight into circuit failure and is of little assistance in debugging a circuit manufacturing process. On-chip implementations of loop-back technique have been demonstrated, [BC04, HBSC05, YE05].

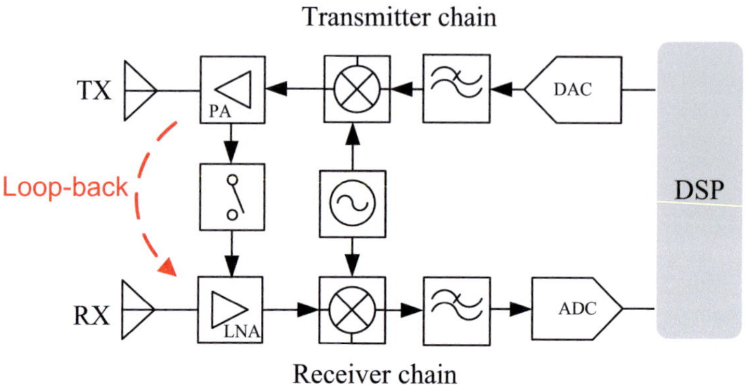

Figure 6.3: Loop-back test concept.

On-chip BIST

The BIST method is one of the most appropriate techniques for testing complex SoCs, as every component in the system can be tested independently from the rest of the system. The main idea behind a BIST approach is to eliminate the need for the external tester by integrating active testing nodes onto the chip. This approach allows applying at-speed tests and eliminates the need for an external tester.

In general BIST provide many advantages in reducing the testing cost. However, it should be pointed out that BIST has some drawbacks as it needs overhead, power and additional circuits. One advantage of BIST is, it can overcome many of the signal quality problems associated with the parasitic effects introduced by cables connecting the equipment to the device. The advantage of BIST can be summarized as:

- no need of fragile probing is necessary. Instead a test socket for fast testing can be used.
- costly external test equipment can be reduced
- parasitic effects introduced by cables connecting the equipment to the device are avoided

- testing speed technology is kept up-to-date with newer-generation integrated circuits
- the built-in sensors provide test access for internal nodes of the whole system leading to better fault diagnosis
- manufacturing tests and sensor outputs can be extended towards on-line monitoring and fault diagnoses of components: a self-testable device can be examined from a remote location
- testing and performance degradation over lifetime can be monitored.

6.3 Porposed BIST methodology

One key challenge to test a highly integrated radar MMIC is the accessibility of individual blocks in the integrated circuit. The ability to control the inputs and to observe the outputs of each building block is necessary for functionality test and monitoring. Potential solutions to enhancing observability include the addition of simple, low-frequency circuitry at internal nodes, whose outputs can be further used for validation and functionality test.

In this section, a BIST methodology for the fast and low-cost monitoring of MMICs using their response to a digitally assisted control voltage or biasing current stimulus is presented. The proposed methodology does not require complex RF instrumentation hence it can be easily implemented on low-cost, and low-frequency testing equipments. The system level description of the radar sensor with BIST features is detailed in Chapter 5.

Several BIST features have been included on the realized chip in order to facilitate simple low-frequency, low-cost testing and on-line monitoring of the circuit performance. The functionality and verification of the BIST features is detailed in Chapter 6. First, divide-by-64 chains have been introduced for both the transmit and receive VCOs. Apart from allowing the VCOs to be locked by external PLLs, the dividers provide a low frequency signal that can be used to independently verify the tuning range and phase noise of the VCOs. Further more, the divider chains allow the VCOs to be locked by an external PLL. The dividers provide a low frequency in terms of 1.8 GHz signal that can be used to independently verify the proper operation and tuning range of the VCOs.

Bidirectional power detectors, capable of measuring the forward and reflected waves, have been inserted between the receive LO distribution tree and the mixers, as well as between the transmitter distribution and the power amplifier, as shown in Figure 5.1. Their role is to monitor the signal power provided by the LO distribution trees and to isolate potential problems either in the mixers, transmit amplifiers, or in the VCOs. Similar detectors were also placed at the outputs of the primary and of the reference channel. Apart from measuring the transmit power, the detector reading of the reflected power can be employed to calculate the reflection coefficient at the TX output and thus determine whether the antenna is well matched. Table 5.1 summarizes the main performance and characteristics of the radar IC.

Furthermore, since the reflected power is monitored, the cause of potentially insufficient signal power can be traced back to improper matching or to inadequate VCO output power. Power detectors were also placed at the PA outputs of both the main and reference channels. Apart from monitoring of the output power, the detector reading of the reflected power can be employed to estimate the reflection coefficient at the transmitter output and thus estimate how well the antenna is matched. A conceptual diagram of the implemented BIST

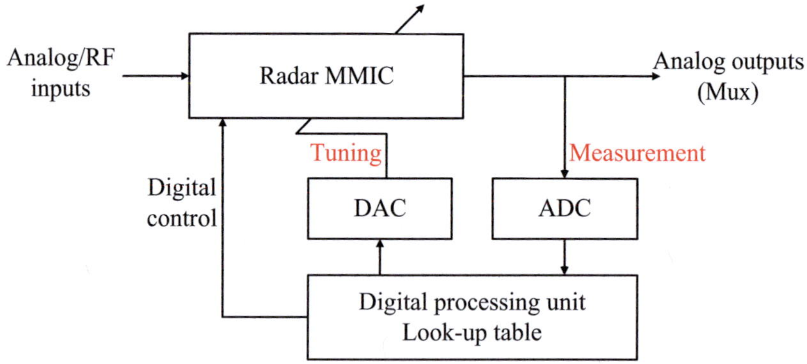

Figure 6.4: Conceptual diagram of a digitally-assisted BIST methodology.

methodology is shown in Figure 6.4. This test concept avoids the use of traditional RF analog testing methods by utilizing low-frequency digitally assisted circuits to autonomously test the integrated components. A digital processing unit (DPU) is used to measure the circuit performance based on the specification values. DAC and ADCs convert the test signals to and from the MMIC, in a low-frequency domain. Internal to the digital part is the DPU responsible for the testing algorithm. The analog block of the MMICs incorporate built-in tuning nodes that can be tuned from the DPU unit for testing function and performance verification. In the DPU unit performance parameters of the blocks can be stored as a look-up table. Table 6.1 highlights the important parameters to test for in an integrated MMIC. Several of these parameters can be extracted using power and amplitude measurements while others require spectral measurements.

Component	**Test specifications**
VCO	output power, tuning range, phase noise
PA	gain, frequency characteristics, output power
LNA	gain, frequency characteristics, output power
Mixer	IF- bandwidth, conversion gain
Divider	functionality
antenna	matching condition

Table 6.1: RF-BIST and monitoring of integrated components

A key components in the BIST system are integrated power detectors at different nodes of the IC and digitally tunable gain functions of the circuit blocks. The presence of these detectors enables the testing and monitoring of the circuitry, thus enables signal tracking along the chain. A digital interface with a 28-bit shift register is integrated in the chip, to monitor and control the individual components. During test configuration, the on-chip analog multiplexer output signal can be read out and compared to the specification values of the components. Thus, the high frequency circuit response is converted into a low frequency form, which is easier to analyze in baseband.

Furthermore since the reflected power is monitored, the cause of potentially insufficient signal power can be traced back to improper matching or to inadequate VCO output power. Another BIST feature in the heterodyne transceiver is the reference channel, which is built symmetrical to the main channel. This architecture allows estimating the complex load at the antenna port, and matching condition of the assembled antenna, as described in chapter 5.

6.4 Component level testing sequence

This section describes the testing sequence of different blocks of the transceiver circuits. To evaluate the functionality and performance of the packaged radar sensor, a test board has been designed and fabricated. The implementation of the test board is based on an architectural focus on a low-frequency BIST function and evaluation. The complete block diagram of the realized circuits and test setup is shown in Figure 6.5. The main components integrated on the PCB are:

- Several power supply components. The radar transceiver operates with fixed 1.2 V and 1.8 V supply voltages, while 2.5 V and 3.3 V supplies are used to supply the DAC, ADC and RF-switch.
- 12-bit programmable DAC (Digital Anlaog Converters): In addition to the fixed supply voltages, several components of the IC have additional biasing supplies, which could be used to adjust and ensure their operation. Furthermore, the coarse tuning of the VCOs could be adjusted by programming the DAC output voltage.
- 12-bit ADC: which is used to digitize the analog multiplex output signal of the radar IC.
- RF-switch: in order to switch between the four RF outputs of the radar IC, a programmable RF-switch is integrated on the test board. The output of the RF-switch is connected to a spectrum analyzer.
- RF-amplifiers: prior to the base-band mixer an RF-amplifier is integrated which is used to compensate the conversion loss of the mixer.
- Baseband mixer: to obtain the baseband signal by mixing the reference and IF signals.
- An SPI interface over a USB (FTDI- Module): is used for all SPI programmable components including the radar transceiver.

The four RF signals (DIV_{TX}, DIV_{RX}, REF and IF) which are coming out of the chip are converted to single ended, by using RF baluns. The single ended signals are then connected to a SP4T RF-switch. A power splitter splits the REF and IF channels, to rout the signal to a mixer and switch, respectively. By multiplexing the control voltages of the switch, one of the RF signal can be displayed at a time on a spectrum analyzer. A miniaturized coaxial connector is used to capture the signal from the RF-output of the switch.

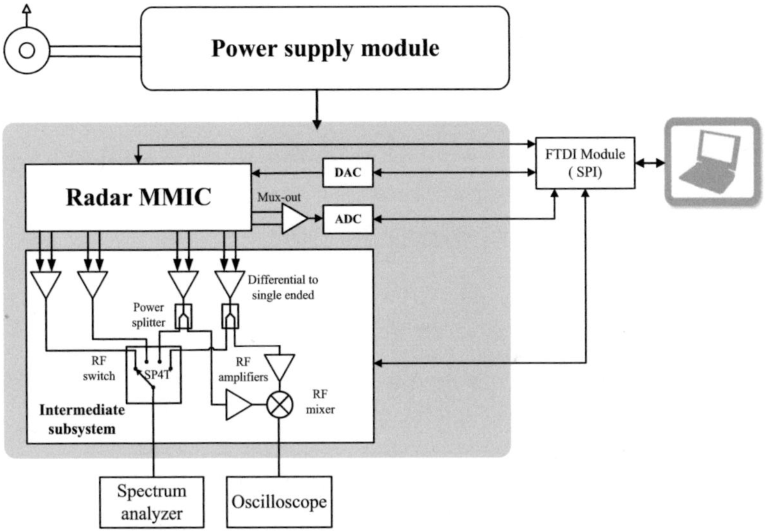

Figure 6.5: System block diagram of the BIST measurement setup. All the RF signals are connected to a miniaturized coaxial connectors. The designed dedicated board is shown in Figure 6.15.

6.4.1 DC test

The total current consumption of the transceiver circuit is 800 mA, from 1.2 V, and 1.8 V supply, when all circuit blocks are active. The power sources and adjustable components (DAC-Voltages) are set to the proper nominal values, as listed in Table 6.2. The circuit is designed in such a fashion that the components could be turned off by setting the bias current sink voltage to 1.2 V

- 1.5 V. Complete turn off happens at 1.8 V. During the DC-test, the circuit blocks could be turned on and off digitally, which in turn is used to check if the power consumption of the components is in the range of the given specification. Catastrophic defects on the packaging (for example a not properly soldered package), or broken down wire-bonds can be easily detected during this DC-test sequence.

Component	**Count**	**Power in [mW]**
VCO	2	76
Divider chain	2	145
LO distribution	1	115
Transmitter chain	1	230
Receiver chain	1	165
Total		**952**

Table 6.2: Power consumption of the MMIC building blocks.

6.4.2 Transmit and receive VCOs

The test parameters to verify the functionality of the VCOs are the tuning characteristics of the VCOs, and the power level of the generated signal. By

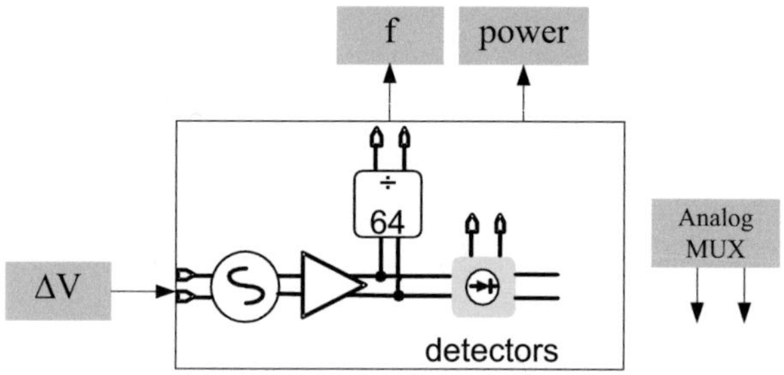

Figure 6.6: VCO testing methodology: the tuning voltage of the VCOs is used to stimulate the VCOs by changing the frequency of operation. To verify the functionality, the frequency can be measured at the divider output and the power level can be measured at the output of the analog multiplexers.

using the test set-up shown in Figure 6.6, it is possible to detect the fault of the

VCO and the divider circuitry. Subsequently, the power detector values were recorded and translated into power by using the measured characterstics of the standalone detectors. The corresponding power for each frequency points is measured independently. The output of the power detectors is connected directly to the analog multiplexer, which are sampled by the integrated on-board ADC. The fail and pass criteria is stored in a look-up table, which is used to

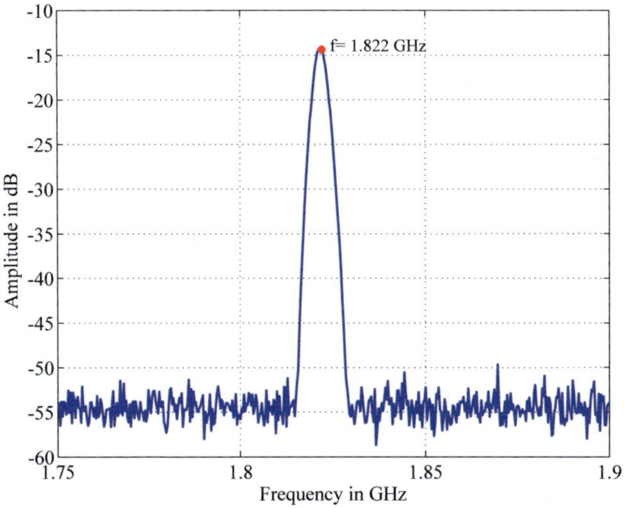

(a) V_{tune}=1.0 V

Figure 6.7: Divider output at different tuning voltages.

compare the DUT with the specified values. Since the transmit and receive VCOs are identical, the same procedure is used in both cases. Note that the signal power level of the divider output signal doesn't correspond to the real power level of the signal generated by the VCOs. Frequency dependent parameters can be measured by successive sweeping of the two on-chip VCOs. The tuning characteristics of the VCOs is shown in Figure 6.8, which can be used as a test specification during the test sequence. In order to obtain a wider bandwidth in FMCW mode of operation, the fine and coarse tuning shall be combined. A typical spectrum outputs of a divider signal for different tuning voltages are shown in Figure 6.7.

A failure of operation on the divider, can be detected at the tuning output. In the case when the divider is defective, the proper operation of the VCO can still be detected by tuning the tuning voltage and capturing the REF and IF signals, which are changing with the tuning voltage of the VCOs. Another way of proving the operation of the VCOs is, by changing the transmit power level of the VCOs digitally, and reading out the detector outputs, in a digital fashion. However, for a radar operation, since the output of the dividers are used to lock the VCOs, a failure of divider operation is categorized as a fully defective DUT. The proper operation of the VCOs was first verified by measuring the frequency at the divider outputs for different tuning voltages, from 117.3 GHz to 127 GHz, as shown in Figure 6.8.

The phase noise has been measured at the divider output as depicted in Figure 6.9. It exhibits a phase-noise of -118 dBc/Hz at 1 MHz offset from the highest oscillation frequency. The signal is in the range of 1.9 GHz and contains the same non-linear tuning curve and phase noise scaled by a factor of 20log(64) dB = 36 dB. Thus, the phase noise value at 122 GHz carrier frequency is -82 dBC/Hz at 1 MHz offset frequency.

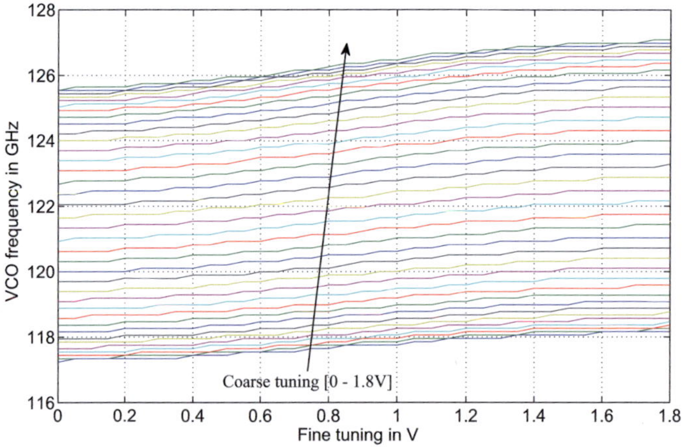

Figure 6.8: Tuning curve of the two VCOs: The coarse and the fine tuning voltages could be tuned in the range of 0 to 1.8 V, which results for the overall frequency span of 117.3 to 127 GHz.

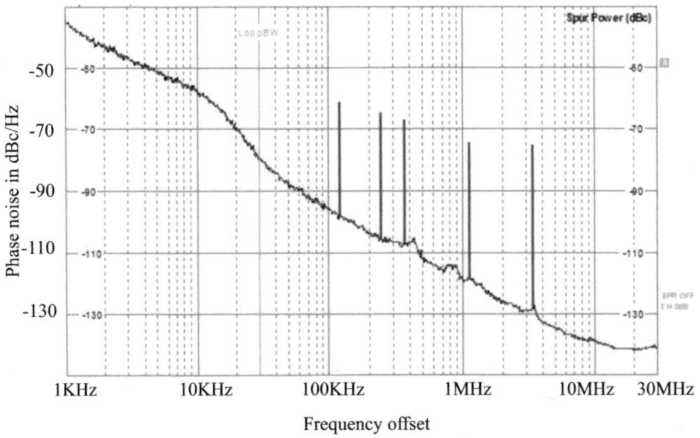

Figure 6.9: Measured phase noise of the radar transceiver at the divider output.

6.4.3 Transmit and receive power amplifiers

The programmable gain in the LNA and PA was realized in a digital circuit by using a 4- bit control circuit for each. The gain level of the two PAs can be tuned digitally, using the integrated serial interface. Their functionality can be therefore verified by reading out the power detector outputs right behind the PAs, or reading out the change in the spectrum's power level at the IF and REF outputs.

The capability to adjust the output power was verified using the on-chip power detectors as well as by monitoring the transmit to receive leakage signal through the 6-dB antenna coupler. The two measurements track each other and indicate that their is over 15 dB of output power control. Figure 6.10 reproduce the analog detector output behavior of the power detectors at 122 GHz.

Another way of testing the power amplifiers, is by changing the gain level and monitoring the IF and REF channel outputs. Any defect on the power amplifers can be easily detected at the outputs of the mixer as shown in Figure 6.11.

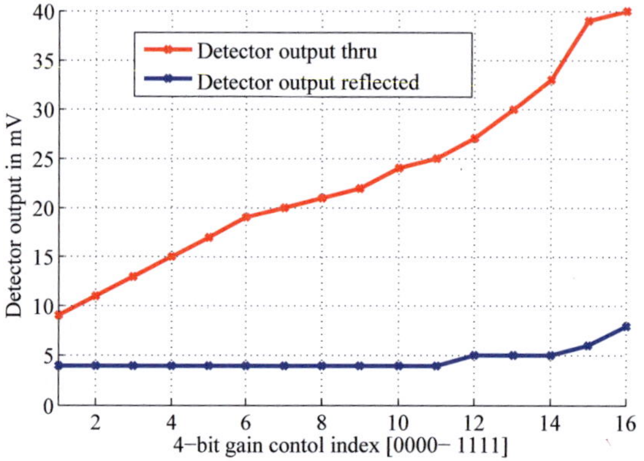

Figure 6.10: Analog output versus detected power characteristics at 122 GHz.

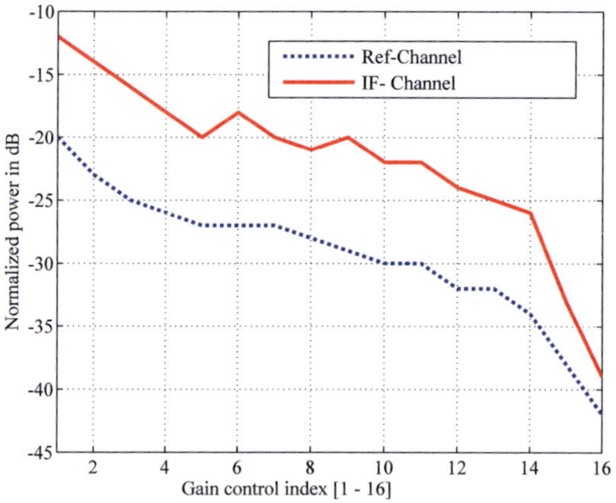

Figure 6.11: Gain control of the LNA versus output power at IF and REF outputs.

6.4.4 Mixer testing of the reference and IF channels

There are a number of ways to test the functionality of the two integrated on-chip mixers. One way of verification is by tuning the two VCOs within the spectrum bandwidth of the mixers, and monitoring the frequency at the IF and REF channel. In this case the $f_{REF} = (f_{TX} \times f_{RX}) \leq 5\,\text{GHz}$. By tuning the LNA gain digitally and reading out the spectrum power at the REF and IF channel verifies the functionality of the mixers, as well. Similar to the LNAs, the RX gain control steps were characterized by employing the transmitter to receiver leakage and measuring the change in the received signal output power, as shown in Figure 6.12. A typical measured normalized IF channel output at $f_{IF} = 476$ MHz is depicted in Figure 6.13. The second peak on the spectrum shows the harmonic value at a frequency of 952 MHz.

In general the following procedures can be performed to test the proper operation of the two integrated mixers:

- Sweep the transmit VCO modulation input, by setting the receive VCO to a fixed frequency. Check REF mixer output power
- Sweep the receive VCO modulation input, while setting the transmit VCO to a fixed frequency. Check receive mixer output power
- Set the transmit VCO modulation input to a fixed value. Sweep the power amplifier power control. Check the output power of the receiver mixer.
- Set the transmit VCO modulation input to a fixed value. Sweep the reference channel power control. Check the output power of the reference mixer.

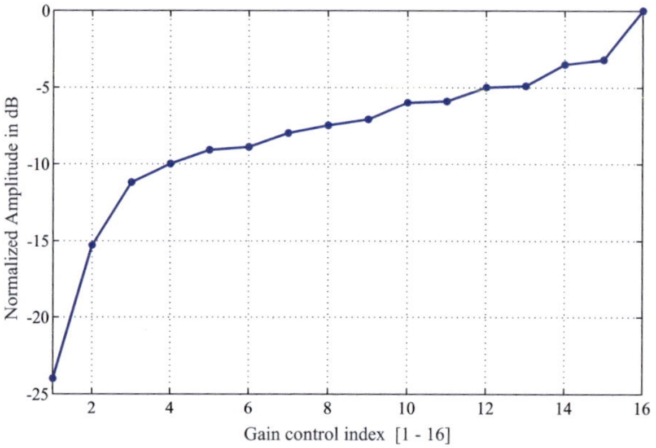

Figure 6.12: Measured IF output versus the gain control of the LNA.

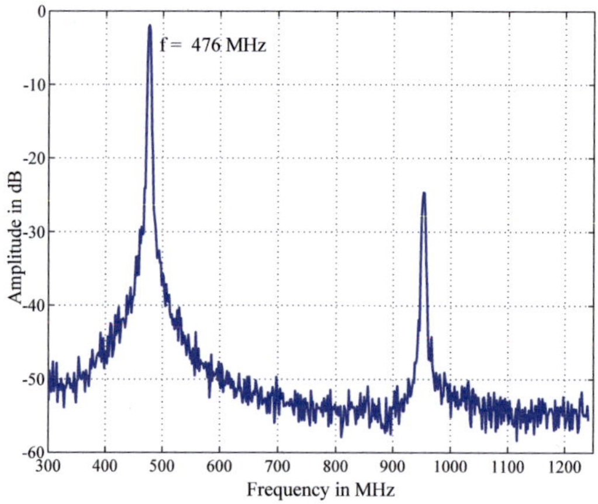

Figure 6.13: Measured IF channel output at f_{IF} = 476 MHz. The second peak on the spectrum shows the harmonic value at a frequency of 952 MHz.

6.5 Antenna match measurement of assembled antenna

Antenna match condition is one of the most critical performance parameters of a highly integrated radar systems. However, the measurement of matching performance of mm-wave antennas is very challenging, after the antenna is integrated in the package. On the radar system architecture described in this chapter, due to the symmetry of the architecture, the on-chip tuner was used to measure the complex reflection coefficients of the antenna connected to the main channel. For this purpose, the main channel was used as the reference and the output of the reference channel was monitored for changes in amplitude and phase. The same procedure described in Section 5.2.2, is used to compute the complex S-parameter of a connected antenna. Any mismatch of the antenna or a defect on the connection of wire-bonds can be detected by using the described methodology.

122 GHz Integrated antenna in package

The planer mm-wave antenna[1] is realized as an antenna array composed of four elements that are fed by a novel feeding network using a grounded Coplanar Waveguide (GCPW), and pairs of coupled microstrip (MS) lines.The antenna is realized on a 0.1 mm thick Rogers Ultralam 3850 substrate, using a thin film technology. Figure 6.14 shows a 122 GHz IC that is integrated into a 8 mm x 8 mm QFN package, together with antenna. The IC and the antenna are connected using three aluminum wire in wedge-wedge technology.

To avoid direct soldering on the test PCB, an open-top test socket with a center opening for package access is used, as shown in Figure 6.15 . In this way, several ICs could be tested in a short time. The block diagram of the implemented test board is shown in Figure 6.5. The S-parameter of the packaged antenna is measured by using the methodology described in section 5.2.2. The comparison of the computed complex reflection coefficients are plotted in Figure 6.16, which shows all the 256 tuning index values of the variable impedance tuner before and after corrections. For similar antennas, a magni-

[1] As a part of the EU project SUCCESS, the mm-wave antenna design and packaging is done by KIT. The details of the antenna design and packaging concept are presented in [BRG+13].

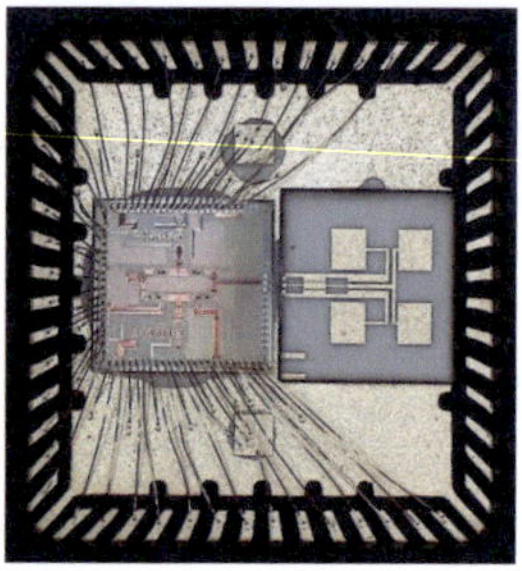

Figure 6.14: 8mm x 8mm QFN package with integrated IC and antenna on alumina substrate.

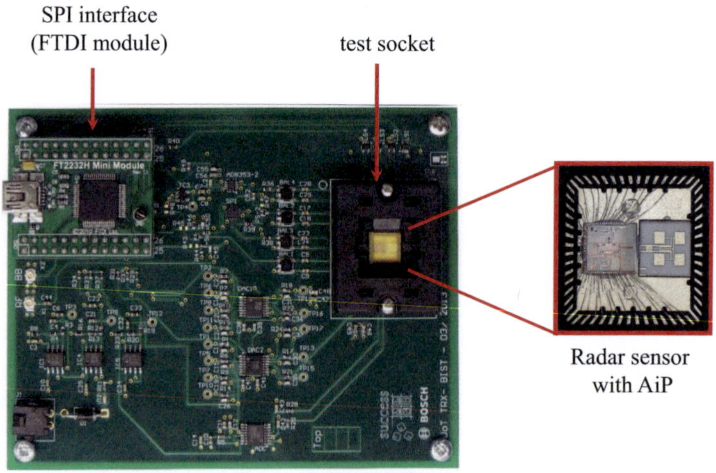

Figure 6.15: Realized PCB with a 56-pin test socket for BIST and verification. The packaged sensor is covered with a dedicated sensor cap, which is designed as an RF- transparent at the center frequency of 122.5 GHz.

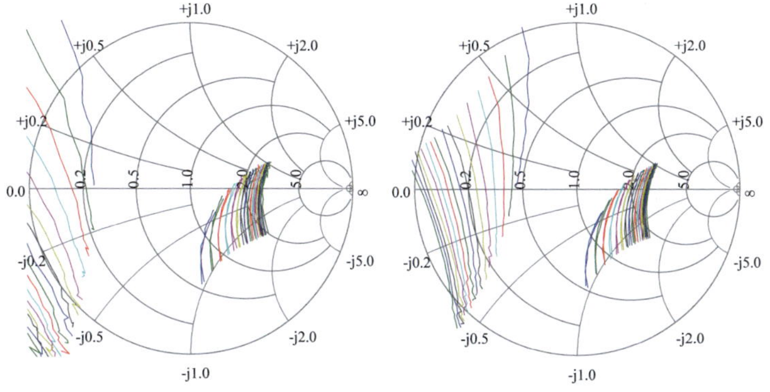

(a) Antenna-1 (b) Antenna-2

Figure 6.16: Complex reflection coefficients of two assembled antennas at 122 GHz for all 256 tuning indexes. The measurements include the influence of the on-chip transmission line, the connecting wire bonds, the cap and the test socket.

tude comparison of the simulation and measurement return losses is plotted in Figure 6.17, which proves the functionality of the proposed methodology.

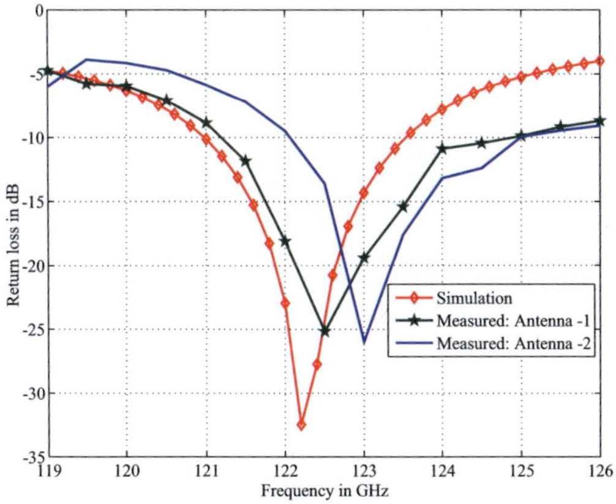

Figure 6.17: Comparison of S11 amplitude simulation and embedded measurements of the integrated antennas after assembly. The measurement of the antennas are done in 500 MHz frequency steps.

6.6 Concluding remarks

An on-chip integrated low-cost BIST scheme for testing of radar transceiver chip is presented in this chapter. Several self-test features have been included in order to facilitate simple low-frequency, low-cost testing and thus minimize or avoid of D-band equipment. This BIST methodology enables the on-line monitoring of all the integrated components in a low-frequency and digital fashion. Furthermore, for the first time it is shown to measure the matching condition of an assembled antenna in mm-wave frequency. A novel technique have been developed and verified to accurately quantify the S-parameter of the integrated mm-wave antennas. The proposed method can be used to measure the assembly or packaging errors introduced by different mounting methods at mm-wave frequencies.

7 Summary and outlook

In this dissertation a technology and system architecture study of mm-wave radar sensors for short range applications has been performed. Emphasis has primarily been on the system concept and integration. The study includes a theoretical analysis of different mm-wave radar architectures, including a dual-channel homodyne as well as a complex heterodyne architecture. A novel BIST concept for mm-wave radar sensors and a embedded calibration methodology is proposed and investigated.

Chapter 2 presents the basics of radar system architecture and requirements on system design by giving emphasis on near range application. Comparison of different system architectures have been briefly investigated, which offered insight into the operation principles of the presented architectures. The state-of the-art millimeter wave antenna integration and packaging methods were extensively reviewed and discussed regarding their advantages and limitations.

Chapter 3 describes the design considerations, integration issues, packaging and experimental performance of a recently developed D-Band, dual-channel radar transceiver with on-chip antennas fabricated in a SiGe-BiCMOS technology. The design comprises of a fully integrated transceiver circuit with quasi-monostatic architecture and a 60 GHz push-push VCO that operates between frequency band of 117.5 GHz and 123.5 GHz. All analog building blocks have digital-control interfaces and are controlled via SPI to reduce the number of connections and facilitate the communication between digital processor and analogue building blocks. The two electromagnetically coupled patch antennas are placed on the top of the die with 9-dBi gain and have a simulated efficiency of 60% .The chip consumes 450 mW and is wire-bonded in an open-lid 5 mm × 5 mm QFN package. Appropriate signal processing for the estimation of range, and azimuth angle in single object situation is presented.

Chapter 4 presented the realization of a novel heterodyne FMCW radar system integrated on a single chip. The principle is the use of two separate signal sources integrated on the same chip and an external down-conversion to correlated the phase noise of the two different sources. High performance ranging accuracy for short distance measurement is shown. A performance comparison of four different sensor architectures is done.

Chapter 5 presents an embedded calibration methodology of a radar sensor system with an integrated reference channel. A digital impedance tuner is used to generate a known complex impedance values, which are later used to estimate the load at the antenna port. A one port VNA methodology with 256 measurements is employed to characterize the error coefficients. Imperfections such as offset, LO leakage, mismatching are thus determined. The calibration method has been verified by measurements, which shows that with the use of the reference channel we achieve improved signal linearity.

In **Chapter 6** it has been shown that a variety of built-in self test schemes are currently the topic of research for both the millimeter wave and mixed-signal domains. Due to aggressive technology scaling and multi-GHz operating frequencies of RF devices, parametric failure test and diagnosis of RF circuitry is becoming increasingly important for the reduction of production test cost and faster yield ramp-up. Methodologies and solutions for testing individual building blocks of an integrated circuits, that will enable direct verification of performance parameters have been presented throughout this chapter. Current RF transceiver architectures will require a mixture of both paradigms for successful implementation of integrated test strategies. The application of BIST concepts is a must for future integrated transceiver systems to meet the stringent requirement of time-to-market and low-cost of mass market consumer products.

A new BIST scheme for testing of a radar transceiver chip with an integrated reference channel is realized and implemented. A low-cost and low frequency test and diagnosis method is presented to allow fault prediction and on-line monitoring of the integrated components in the radar transceiver. This low-frequency test methodology provide the features for testing without external RF ATE support. For prove of concept a single chip, heterodyne monostatic radar architecture with a BIST capability is realized and investigate. Several self-test features have been included in order to facilitate simple testing methodology and thus minimize or avoid of D-band equipment. Furthermore, for the first time it is shown to measure the matching condition of an assembled antenna in mm-wave frequency. A novel technique have been developed and verified to accurately quantify the S-parameter of the integrated mm-wave antennas. Within the framework of this PhD work, the following novel results and insights have been achieved:

- Miniaturized dual-channel radar sensor at 122 GHz with integrated on-chip antennas for distance and angle measurements was developed. The transceiver is based on a SiGe-BiCMOS chip with two separate integrated antennas. The SMD radar requires no external millimeter wave connections, thus enabling the assembly in a standard SMD line. Due to the two receiving channels, an angular measurement in azimuth direction is possible. These results demonstrate the potential of a compact, low-cost D-band radar sensor for a consumer application.

- In general, the DC-offset in homodyne radar architectures is more severe than heterodyne architectures, thus the former systems are not appropriate for a short range measurements. However, heterodyne receivers are more complex, since more signal sources and down conversion stages are required. In spite of these disadvantages the heterodyne radar system has shown a better performance for short range measurements. In a case study, it is shown that with a calibration (empty signal subtraction) of a radar signal prior to the measurements, it will not be fully possible to overcome the dynamic DC-offset which usually varies fast over the radar front-end system temperature.

- In this work, for the first time in the literature, the potential of a heterodyne radar system architecture for accurate near range detection is investigated. A novel single chip heterodyne FMCW system with two separate sweeping sources and a demodulation scheme is used. This architectures mitigates the effect of DC-offsets in the near range. High distance measurement accuracy at a short distance below 50 cm have been achieved compared to a homodyne architecture, which suffers from a DC-offset issue.

- A novel, embedded auto-calibration methodology of system impairments in integrated radar sensors is shown for the first time. For the calibration reference channel a digital impedance tuner with a look-up table over a frequency is used. The calibration method can adequately calibrate non-idealities and systematic errors in radar sensors in analog domain.

- A BIST concept and methodology for highly integrated radar sensors has been proposed. The proposed approach allows a straightforward low frequency and digital testing of each individual component integrated on chip. All the circuitry succeeding the BIST will simply convert the sensed information into a low-frequency equivalent and post-process that signal. The methodology enables an online monitoring of the sensor system during operation.
- On-chip level built-in testing methodology to measure the complex S_{11} of an integrated antenna is demonstrated for the first time. This novel technique has been developed and verified to accurately quantify the match condition of integrated mm-wave antennas. Lower cost and improved reliability can be potentially achieved by on-line monitoring of antenna match condition. This can potentially replace the classical on-wafer probe testing method or expensive probe cards can drastically reduce the testing cost and enable new test capabilities. Another key advantage of using a built-in test methodology is, that an on-line monitoring of the circuits is possible after assembly.

In this PhD thesis, several mm-wave radar systems with emphasis on short range distance measurements are presented and there potential to replace the state-of-the-art measuring systems is verified. System architecture concepts for performance improvement, for example the short range measurement capability of the sensor, as well as cost-effective testing methodologies are introduced. Note that all of these developed concepts could be transferred to other application areas, such as automotive radar systems at 77 GHz frequency spectrum.

Outlook

One major research topic that would complement this work is a development of a fully functional FMCW radar system based on the radar sensor system architecture which is shown in Chapter 5. This could verify the advantage of the embedded systematic error calibrations on the performance of the sensor system. In addition, the potential of the two channel radar system to measure a complex S-parameter can be exploit to develop a one-port VNA system for applications, such as material characterization.

Own publications and patents

[1] M.G. Girma, J. Hasch, I. Sarkas, S.P. Voinigescu, and T. Zwick, 122 GHz radar sensor based on a monostatic SiGe-BiCMOS IC with an on-chip antenna. The *7th European Microwave Integrated Circuits Conference (EuMIC)*, pages 357–360, October 2012.

[2] M.G. Girma, S. Beer, J. Hasch, M. Gonser, W. Debski, W. Winkler, Yaoming Sun, and T. Zwick, Miniaturized 122 GHz system-in-package (SiP) short range radar sensor. In *European Radar Conference (EuRAD)*, pages 49–52, October 2013.

[3] S. Beer, M.G. Girma, S. Yaoming, W. Winkler, W. Debski, J. Paaso, G. Kunkel, J.C. Scheytt, J. Hasch, and T. Zwick, Flip-chip package with integrated antenna on a polyimide substrate for a 122-GHz bistatic radar IC. The *7th European Conference onAntennas and Propagation (EuCAP)*, pages 121–125, April 2013.

[4] S. Beer, C. Rusch, H. Gulan, B. Göttel, M.G. Girma, J. Hasch, W. Winkler, W. Debski, and T. Zwick. An Integrated 122-GHz Antenna Array With Wire Bond Compensation for SMT Radar Sensors, *IEEE Transactions on Antennas and Propagation*, 61(12), pages 5976–5983, December 2013.

[5] B. Göttel, M. Pauli, H. Gulan, M. Girma, J. Hasch, and T. Zwick, Miniaturized 122 GHz short range radar sensor with antenna-in-package (AiP) and dielectric lens. *The 8th European Conference on Antennas and Propagation (EuCAP)*, pages 709–713, April 2014.

[6] M.G. Girma, M Gonser, A. Frischen, J. Hasch, Y. Sun, and T. Zwick. 122 GHz Single-Chip Dual-Channel SMD Radar Sensor with On-chip Antennas for Distance and Angle Measurements, *International Journal of Microwave and Wireless Technologies*, 7(3-4), pages 407–414, June 2015.

[7] M.G. Girma, J. Hasch, M. Gonser, Yaoming Sun, and T. Zwick. 122 GHz Single-Chip Dual-Channel SMD Radar Sensor with Integrated Antennas for Distance and Angle Measurements, In *European Radar Conference (EuRAD), 2014*, pages 1754 – 1757, October 2014.

[8] I. Sarkas, M.G. Girma, J. Hasch, T. Zwick, and S.P. Voinigescu. A Fundamental Frequency 143-152 GHz Radar Transceiver with Built-In Calibration and Self-Test. In *Compound Semiconductor Integrated Circuit Symposium (CSICS), 2012 IEEE*, pages 1–4, October 2012.

[9] Y. Sun, M. Marinkovic, G. Fischer, W. Winkler, W. Debski, S. Beer, T. Zwick, M.G. Girma, J. Hasch, and C.J. Scheytt. A low-cost miniature 120 GHz SiP FMCW/CW radar sensor with software linearization. In *IEEE International Solid-State Circuits Conference Digest of Technical Papers (ISSCC)*, pages 148–149, February 2013.

[10] A. Frischen, J. Hasch, D. Jetty, M.G. Girma, M. Gonser, and C. Waldschmidt. A Low-Phase-Noise 122-GHz FMCW Radar Sensor for Distributed Networks, In *European Radar Conference (EuRAD)*, pages 149 –52, October 2016.

[11] G. Hakobyan, M. Girma, X. Li, N. Tammireddy, and B. Yang. Repeated Symbols OFDM-MIMO Radar at 24 GHz, In *European Radar Conference (EuRAD)*, pages 249 – 252, October 2016.

[12] S. Shopov, M.G. Girma, J. Hasch, and S.P.Voinigescu. An Ultra-Low-Power 4-Channel 60-GHz Radar Sensor, *IEEE MTT-S International Microwave Symposium (IMS)*, pages 1520–1523, June 2017.

[13] S. Shopov, M.G. Girma, J. Hasch, N. Cahoon and S.P.Voinigescu. Ultra-Low-Power Radar Sensors for Ambient Sensing in the V-Band, *IEEE Transaction on Microwave Theory and Techniques*, pages 5401–5410, December 2017.

[14] M.G. Girma, J. Hasch, and W. Soergel. Method for calibrating a radar system, *US Patent App. 15535321*, November 2017.

[15] J. Meyer, S. Bannwarth, W. Hafla, M.G. Girma, J. Hasch, T. Binzer, W. Soergel, C. Potratz. Antenna device for receiving electromagnetic waves and method for operating an antenna device for receiving electromagnetic waves, *US Patent App. 15174636*, December 2016.

Bibliography

[66198] IEEE Standard Radar Definitions. *IEEE Std 686-1997*, pages 1–39, 1998.

[ADC+09] G. Avenier, M. Diop, P. Chevalier, G. Troillard, N. Loubet, J. Bouvier, L. Depoyan, N. Derrier, M. Buczko, C. Leyris, S. Boret, S. Montusclat, A. Margain, S. Pruvost, S. T. Nicolson, K. H. K. Yau, N. Revil, D. Gloria, D. Dutartre, S. P. Voinigescu, and A. Chantre. 0.13 um SiGe BiCMOS Technology Fully Dedicated to mm-Wave Applications. *IEEE Journal of Solid-State Circuits*, 44(9):2312–2321, September 2009.

[AGW+11] Morteza Abbasi, S.E. Gunnarsson, N. Wadefalk, R. Kozhuharov, J. Svedin, S. Cherednichenko, I. Angelov, I. Kallfass, P. Leuther, and H. Zirath. Single-Chip 220-GHz Active Heterodyne Receiver and Transmitter MMICs With On-Chip Integrated Antenna. In *IEEE Transactions on Microwave Theory and Techniques*, volume 59, pages 466–478, February 2011.

[AHS11] M. Al Henawy and M. Schneider. Integrated antennas in eWLB packages for 77 GHz and 79 GHz automotive radar sensors. In *41st European Microwave Conference (EuMC)*, pages 1312–1315, October 2011.

[AO05] E. Acar and S. Ozev. Defect-based RF testing using a new catastrophic fault model. In *IEEE International Test Conference (ITC) Proceedings*, pages 420–429, Nov 2005.

[Arm24] E. H. Armstrong. The Super-Heterodyne-Its Origin, Development, and Some Recent Improvements. *Proceedings of the Institute of Radio Engineers*, 12(5):539–552, October 1924.

[AS07] S. Abdennadher and S.A. Shaikh. Practices in Mixed-Signal and RF IC Testing. *IEEE Design Test of Computers*, 24(4):332–339, July 2007.

[ASG+12] S. S. Ahmed, A. Schiessl, F. Gumbmann, M. Tiebout, S. Methfessel, and L. P. Schmidt. Advanced Microwave Imaging. *IEEE Microwave Magazine*, 13(6):26–43, September 2012.

[BB93] M. C. Budge and M. P. Burt. Range correlation effects on phase and amplitude noise. In *IEEE Southeastcon '93, Proceedings.*, pages 5 –9, April 1993.

[BC04] S. Bhattacharya and A. Chatterjee. A built-in loopback test methodology for RF transceiver circuits using embedded sensor circuits. In *13th Asian Test Symposium*, pages 68–73, Nov 2004.

[Bee13] Stefan Beer. *Methoden und Techniken zur Integration von 122 GHz Antennen in miniaturisierte Radarsensoren.* PhD thesis, Karlsruher Institut of Technologie (KIT), 2013.

[Bel02] J.S. Belrose. Reginald Aubrey Fessenden and the birth of wireless telephony. *IEEE Antennas and Propagation Magazine*, 44(2):38–47, 2002.

[BFBM06] M. Brunnbauer, E. Furgut, G. Beer, and T. Meyer. Embedded wafer level ball grid array (eWLB). In *8th Electronics Packaging Technology Conference, EPTC*, pages 1–5, December 2006.

[BGK+06] A. Babakhani, X. Guan, A. Komijani, A. Natarajan, and A. Hajimiri. A 77-GHz Phased-Array Transceiver With On-Chip Antennas in Silicon: Receiver and Antennas. *IEEE Journal of Solid-State Circuits*, 41(12):2795–2806, December 2006.

[BGRZ12a] S. Beer, H. Gulan, C. Rusch, and T. Zwick. Coplanar 122-GHz Antenna Array With Air Cavity Reflector for Integration in Plastic Packages. *IEEE, Antennas and Wireless Propagation Letters*, pages 160–163, 2012.

[BGRZ12b] S. Beer, H. Gulan, G. Rusch, C. Kunkel, and T. Zwick. 122-GHz Chip-to-Antenna Wire Bond Interconnect with High Repeatability. *IEEE MTT-S International Microwave Symposium Digest*, pages 1–3, 2012.

[BGRZ13] S. Beer, H. Gulan, C. Rusch, and T. Zwick. Integrated 122-GHz Antenna on a Flexible Polyimide Substrate With Flip Chip Interconnect. In *IEEE Transactions on Antennas and Propagation*, volume 61, pages 1564–1572, April 2013.

[BGY+13] S. Beer, M.G. Girma, S. Yaoming, W. Winkler, W. Debski, J. Paaso, G. Kunkel, J.C. Scheytt, J. Hasch, and T. Zwick. Flip-chip package with integrated antenna on a polyimide substrate for a 122-GHz bistatic radar IC. In *7th European Conference on Antennas and Propagation (EuCAP)*, pages 121–125, April 2013.

[BL15] J. Bock and R. Lachner. SiGe BiCMOS and eWLB packaging technologies for automotive radar solutions. In *IEEE MTT-S International Conference on Microwaves for Intelligent Mobility (ICMIM)*, pages 1–4, April 2015.

[BRD+11] S. Beer, B. Ripka, S. Diebold, H. Gulan, C. Rusch, P. Pahl, and T. Zwick. Design and Measurement of Matched Wire Bond and Flip Chip Interconnects for D-Band System-in-Package. *IEEE International Microwave Symposium Digest (MTT-S)*, pages 1–4, 2011.

[BRG+13] S. Beer, C. Rusch, H. Gulan, B. Gottel, M.G. Girma, J. Hasch, W. Winkler, W. Debski, and T. Zwick. An Integrated 122-GHz Antenna Array With Wire Bond Compensation for SMT Radar Sensors. *IEEE Transactions on Antennas and Propagation*, 61(12):5976–5983, December 2013.

[BSA+03] A.R. Behzad, Zhong Ming Shi, S.B. Anand, Li Lin, K.A. Carter, M.S. Kappes, Tsung-Hsien Lin, T. Nguyen, D. Yuan, S. Wu, Y.C. Wong, Victor Fong, and A. Rofougaran. A 5-GHz direct-conversion CMOS transceiver utilizing automatic frequency control for the IEEE 802.11a wireless LAN standard. *IEEE Journal of Solid-State Circuits*, 38(12):2209–2220, December 2003.

[BSEdO05] I. Bashir, R.B. Staszewski, O. Eliezer, and E. de Obaldia. Built-in self testing (BIST) of RF performance in a system-on-chip (SoC). In *Proceedings of Architecture, Circuits and Implementtation of SOCs,*, pages 215–218, 2005.

[CC10] K.-T. Cheng and H.-M Chang. Recent Advances in Analog, Mixed-Signal, and RF Testing. *IPSJ Transcations on System LSI Design Methodology (TSLDM)*, 3(5):19–46, February 2010.

[CCYH04] Tie Jun Cui, Weng Cho Chew, Xiao Xing Yin, and Wei Hong. Study of resolution and super resolution in electromagnetic imaging for half-space problems. *IEEE Transactions on Antennas and Propagation*, 52(6):1398–1411, June 2004.

[CDC+08] K. B. Cooper, R. J. Dengler, G. Chattopadhyay, E. Schlecht, J. Gill, A. Skalare, I. Mehdi, and P. H. Siegel. A High-Resolution Imaging Radar at 580 GHz. *IEEE Microwave and Wireless Components Letters*, 18(1):64–66, January 2008.

[CDL+11] K. B. Cooper, R. J. Dengler, N. Llombart, B. Thomas, G. Chattopadhyay, and P. H. Siegel. Thz imaging radar for standoff personnel screening. *IEEE Transactions on Terahertz Science and Technology*, 1(1):169–182, September 2011.

[CDN75] M. Cohn, James E. Degenford, and B.A. Newman. Harmonic Mixing with an Antiparallel Diode Pair. In *IEEE Transactions on Microwave Theory and Techniques*, volume 23, pages 667–673, August 1975.

[CGH+08] Chang-Soon Choi, E. Grass, F. Herzel, M. Piz, K. Schmalz, Yaoming Sun, S. Glisic, M. Krstic, K. Tittelbach, M. Ehrig, W. Winkler, R. Kraemer, and C. Scheytt. 60GHz OFDM Hardware Demonstrators in SiGe BiCMOS: State-of-the-art and future development. In *IEEE 19th International Symposium on Personal, Indoor and Mobile Radio Communications, PIMRC.*, pages 1–5, September 2008.

[Cha00] K. Chang. *RF and Microwave Wireless Systems*. Wiley, 2000.

[CLJ+14] Ka Fai Chang, Rui Li, Cheng Jin, Teck Guan Lim, Soon Wee Ho, How Yuan Hwang, and Boyu Zheng. 77-GHz Automotive Radar Sensor System With Antenna Integrated Package. *IEEE Transactions on Components, Packaging and Manufacturing Technology*, 4(2):352–359, February 2014.

[CPL+09] P. Chevalier, F. Pourchon, T. Lacave, G. Avenier, Y. Campidelli, L. Depoyan, G. Troillard, M. Buczko, D. Gloria, D. Celi, C. Gaquiere, and A. Chantre. A Conventional Double-Polysilicon FSA-SEG Si/SiGe: C HBT reaching 400 GHz f_{MAX}. In *IEEE Bipolar BiCMOS Circuits and Technology Meeting, BCTM*, pages 1–4, October 2009.

[DKW10] L. Maurer D. Kissinger, B. Laemmle and R. Weigel. Integrated Test for Silicon Front Ends. In *IEEE Microwave Magazine*, pages 87–94, May 2010.

[DR11] N. Deferm and P. Reynaert. A 120GHz 10Gb/s phase-modulating transmitter in 65nm LP CMOS. In *IEEE International Solid-State Circuits Conference Digest of Technical Papers (ISSCC)*, pages 290–292, February 2011.

[DSIP13] Manh-Tuan Dao, Dong-Hun Shin, Yun-Taek Im, and Seong-Ook Park. A Two Sweeping VCO Source for Heterodyne FMCW Radar. *IEEE Transactions on Instrumentation and Measurement*, 62(1):230–239, 2013.

[DWGK11] W. Debski, W. Winkler, D. Genschow, and R. Kraemer. 24 GHz transceiver front-end with integrated ramp generator. In *European Microwave Integrated Circuits Conference (EuMIC)*, pages 21–24, October 2011.

[DWS+12] W. Debski, W. Winkler, Y. Sun, Marinkovic M., J. Borngraeber, and C. Scheytt. 120 GHz Radar Mixed-Signal Transceiver. In *proc. 7th European Microwave Integrated Circuits Conference (EuMiC)*, pages 191–194, 2012.

[EWA+11] S. Emami, R.F. Wiser, E. Ali, M.G. Forbes, M.Q. Gordon, Xiang Guan, S. Lo, P.T. McElwee, J. Parker, J.R. Tani, J.M. Gilbert, and C.H. Doan. A 60GHz CMOS phased-array transceiver pair for multi-Gb/s wireless communications. In *IEEE International Solid-State Circuits Conference Digest of Technical Papers (ISSCC)*, pages 164–166, February 2011.

[FKJ+08] H. P Forstner, H. Knapp, H. Jager, E. Kolmhofer, J. Platz, F. Starzer, M. Treml, A. Schinko, G. Birschkus, J. Bock, K. Aufinger, R. Lachner, T. Meister, H. Schafer, D. Lukashevich, S. Boguth, A. Fischer, F. Reininger, L. Maurer, J. Minichshofer, and D. Steinbuch. A 77 GHz 4-channel automotive radar transceiver in SiGe. In *IEEE Radio Frequency Integrated Circuits Symposium (RFIC).*, pages 233–236, June 2008.

[FKS+11] R. Feger, E. Kolmhofer, F. Starzer, F. Wiesinger, S. Scheiblhofer, and A. Stelzer. A heterodyne 77-GHz FMCW radar with offset PLL frequency stabilization. In *2011 IEEE Topical Conference on Wireless Sensors and Sensor Networks*, pages 9–12, January 2011.

[FMTF11] R. Fujimoto, M. Motoyoshi, K. Takano, and M. Fujishima. A 120 GHz/140 GHz dual-channel ASK receiver using standard 65 nm CMOS technology. In *41st European Microwave Conference (EuMC),*, pages 1189–1192, October 2011.

[FWS11] R. Feger, C. Wagner, and A. Stelzer. An IQ-modulator based heterodyne 77-GHz FMCW radar. In *German Microwave Conference (GeMIC)*, pages 1–4, 2011.

[GBH+13] M.G. Girma, S. Beer, J. Hasch, M. Gonser, W. Debski, W. Winkler, Yaoming Sun, and T. Zwick. Miniaturized 122 GHz system-in-package (SiP) short range radar sensor. In *European Radar Conference (EuRAD)*, pages 49–52, October 2013.

[Gen86] C. Gentili. *Microwave Amplifiers and Oscillaters*. 1986.

[GGF+15] M. G. Girma, M. Gonser, A. Frischen, J. Hasch, Y. Sun, and T. Zwick. 122 GHz single-chip dual-channel SMD radar sensor with on-chip antennas for distance and angle measurements. *International Journal of Microwave and Wireless Technologies*, 7(3-4):407–414, June 2015.

[GHG+14] M.G. Girma, J. Hasch, M. Gonser, Yaoming Sun, and T. Zwick. 122 GHz Single-Chip Dual-Channel SMD Radar Sensor with Integrated Antennas for Distance and Angle Measurements. In *European Radar Conference (EuRAD)*, pages 451–454, October 2014.

[GHP+07] E. Grass, F. Herzel, M. Piz, K. Schmalz, Yaoming Sun, S. Glisic, M. Krstic, K. Tittelbach, M. Ehrig, W. Winkler, C. Scheytt, and R. Kraemer. 60 GHz SiGe-BiCMOS Radio for OFDM Transmission. In *IEEE International Symposiumon Circuits and Systems ISCAS*, pages 1979–1982, May 2007.

[GHS+12] M.G. Girma, J. Hasch, I. Sarkas, S.P. Voinigescu, and T. Zwick. 122 GHz Radar Sensor Based on a Monostatic SiGe-BiCMOS IC with an On-chip Antenna. In *7th European Microwave Integrated Circuits Conference (EuMIC)*, pages 357–360, October 2012.

[GPG+14] B. Göttel, M. Pauli, H. Gulan, M. Girma, J. Hasch, and T. Zwick. Miniaturized 122 GHz short range radar sensor with antenna-in-package (AiP) and dielectric lens. In *The 8th European Conference on Antennas and Propagation (EuCAP)*, pages 709–713, April 2014.

[GSH+11] S. Glisic, K. Schmalz, F. Herzel, Rouyu Wang, M. Elkhouly, Yaoming Sun, and J.C. Scheytt. A fully integrated 60 GHz transmitter front-end in SiGe BiCMOS technology. In *IEEE 11th Topical Meeting on Silicon Monolithic Integrated Circuits in RF Systems (SiRF)*, pages 149–152, January 2011.

[Gue18] Oliver Guenther. *Modellierung und Leakage-Kompensation von 77 GHz FMCW-Weitbeeichsradar Transceivern in SiGe-Technologie für KFZ-Anwendungen*. PhD thesis, University of Erlangen- Nuernberg, 2018.

[HBC07] D. Han, S. Bhattacharya, and A. Chatterjee. Low-cost parametric test and diagnosis of RF systems using multi-tone response envelope detection. *Computers Digital Techniques, IET*, 1(3):170–179, May 2007.

[HBSC05] A. Halder, S. Bhattacharya, G. Srinivasan, and A. Chatterjee. A system-level alternate test approach for specification test of RF transceivers in loopback mode. In *18th International Conference on VLSI Design*, pages 289–294, January 2005.

[HJL87] Donald H. Layton Hans J. Liebe. Millimeter-Wave Properties of the Atmosphere: Laboratory Studies and Propagation Modeling. Technical report, October, 1987.

[HSB$^+$16] M. Hitzler, S. Saulig, L. Boehm, W. Mayer, W. Winkler, and C. Waldschmidt. Compact bistatic 160 GHz transceiver MMIC with phase noise optimized synthesizer for FMCW radar. In *2016 IEEE MTT-S International Microwave Symposium (IMS)*, pages 1–4, May 2016.

[HSB$^+$17] M. Hitzler, S. Saulig, L. Boehm, W. Mayer, W. Winkler, N. Uddin, and C. Waldschmidt. Ultracompact 160-GHz FMCW Radar MMIC With Fully Integrated Offset Synthesizer. *IEEE Transactions on Microwave Theory and Techniques*, 65(5):1682–1691, May 2017.

[HTC96] C.D. Hull, J.L. Tham, and R.R. Chu. A direct-conversion receiver for 900 MHz (ISM band) spread-spectrum digital cordless telephone. *IEEE Journal of Solid-State Circuits,*, 31(12):1955–1963, December 1996.

[HTS$^+$12] J. Hasch, E. Topak, R. Schnabel, T. Zwick, R. Weigel, and C. Waldschmidt. Millimeter-Wave Technology for Automotive Radar Sensors in the 77 GHz Frequency Band. *IEEE Transactions on Microwave Theory and Techniques*, 60(3):845 –860, March 2012.

[HWGH10] J. Hasch, U. Wostradowski, S. Gaier, and T. Hansen. 77 GHz radar transceiver with dual integrated antenna elements. In *German Microwave Conference*, pages 280 –283, March 2010.

[HWHM11] J. Hasch, U. Wostradowski, R. Hellinger, and D. Mittelstrass. 77 GHz automotive radar sensor in low-cost PCB technology. In *European Radar Conference (EuRAD)*, pages 101–104, 2011.

[HWS+07] M. Hartmann, C. Wagner, K. Seemann, J. Platz, H. Jager, and R. Weigel. A Low-Power Low-Noise Single-Chip Receiver Front-End for Automotive Radar at 77 GHz in Silicon-Germanium Bipolar Technology. In *IEEE Radio Frequency Integrated Circuits Symposium (RFIC)*, pages 149–152, June 2007.

[IEE93] IEEE. IEEE Symposium on Consumer Applications of Radar and Sonar. 1993.

[IGKR13] O. Inac, F. Golcuk, T. Kanar, and G.M. Rebeiz. A 90- 100-GHz Phased-Array Transmit/Receive Silicon RFIC Module With Built-In Self-Test. *IEEE Transactions on Microwave Theory and Techniques,*, 61(10):3774–3782, 2013.

[Inf15] Infineon. Infinoen - http://www.infineon.com/. visited on December, 2015.

[Jac86] M. C. Jackson. The geometry of bistatic radar systems. *Communications, Radar and Signal Processing, IEE Proceedings F*, 133(7):604–612, December 1986.

[JHTS11] M. Jahn, A. Hamidipour, Z. Tong, and A. Stelzer. A 120 GHz FMCW Radar Frontend Demonstrator Based on a SiGe Chipset. In *proc. 41st European Microwave Conference (EuMC)*, pages 519–522, 2011.

[JKS11] Martin Jahn, H. Knapp, and A. Stelzer. A 122 GHz SiGe-Based Signal-Generation Chip Employing a Fundamental-Wave Oscillator With Capacitive Feedback Frequency-Enhancement. In *IEEE Journal of Solid-State Circuits,*, volume 46, pages 2009–2020, September 2011.

[JSH10] M. Jahn, A. Stelzer, and A. Hamidipour. Highly integrated 79, 94, and 120-GHz SiGe radar frontends. In *IEEE International Microwave Symposium Digest (MTT)*, pages 1324–1327, May 2010.

[KGK+10] S. Koch, M. Guthoerl, I. Kallfass, P. Leuther, and S. Saito. A 120 - 145 GHz Heterodyne Receiver Chipset Utilizing the 140 GHz Atmospheric Window for Passive Millimeter-Wave Imaging Applications. volume 45, pages 1961–1967, October 2010.

[Kis12] D. Kissinger. *Millimeter-Wave Receiver Concepts for 77 GHz Automotive Radar in Silicon-Germanium Technology*. Springer, 2012.

[Lau02] Wai Yuen Lau. Measurement challenges for on-wafer RF-SOC test. In *27th Annual IEEE/SEMI International Electronics Manufacturing Technology Symposium*, pages 353–359, 2002.

[LCSV09] E. Laskin, P. Chevalier, B. Sautreuil, and S.P. Voinigescu. A 140-GHz double-sideband transceiver with amplitude and frequency modulation operating over a few meters. In *IEEE Bipolar/BiCMOS Circuits and Technology Meeting (BCTM)*, pages 178–181, October 2009.

[LDH+13] H. V. Le, H. T. Duong, A. T. Huynh, R. J. Evans, and E. Skafidas. A CMOS 77 GHz radar receiver front-end. In *European Microwave Integrated Circuits Conference (EuMIC)*, pages 13–16, October 2013.

[LLHH10a] J. Lee, Y. A. Li, M. H. Hung, and S. J. Huang. A fully-integrated 77-ghz fmcw radar transceiver in 65-nm cmos technology. *IEEE Journal of Solid-State Circuits*, 45(12):2746–2756, December 2010.

[LLHH10b] Jri Lee, Yi-An Li, Meng-Hsiung Hung, and Shih-Jou Huang. A Fully-Integrated 77-GHz FMCW Radar Transceiver in 65-nm CMOS Technology. volume 45, pages 2746–2756, December 2010.

[LMW04] Kaihui Lin, R.H. Messerian, and Yuanxun Wang. A digital leakage cancellation scheme for monostatic FMCW radar. In *IEEE International Microwave Symposium Digest (MTT-S)*, volume 2, pages 747–750 Vol.2, 2004.

[LWPS06] K. Lin, Y. E. Wang, C. K. Pao, and Y. C. Shih. A ka-band fmcw radar front-end with adaptive leakage cancellation. *IEEE Transactions on Microwave Theory and Techniques*, 54(12):4041–4048, December 2006.

[MOH+10] T. Mitomo, N. Ono, H. Hoshino, Y. Yoshihara, O. Watanabe, and I. Seto. A 77 GHz 90 nm CMOS Transceiver for FMCW Radar Applications. volume 45, pages 928–937, April 2010.

[MS01] T. Musch and B. Schiek. Measurement of the ramp linearity of extremely linear frequency ramps using a fractional dual loop structure. *IEEE Transactions on Instrumentation and Measurement*, 50(2):389–392, April 2001.

[MSW+10] D. Mittelstrass, D. Steinbuch, T. Walter, C. Waldschmidt, and R. Weigel. DC-offset compensation in a bistatic 77GHz-FMCW-radar. In *European Radar Conference (EuRAD)*, pages 447–450, September 2010.

[NAFBSN10] M. R. Nezhad-Ahmadi, M. Fakharzadeh, B. Biglarbegian, and S. Safavi-Naeini. High-efficiency on-chip dielectric resonator antenna for mm-wave transceivers. *IEEE Transactions on Antennas and Propagation*, 58(10):3388–3392, October 2010.

[NHNA+12] M. Neshat, D. M. Hailu, M. R. Nezhad-Ahmadi, G. Z. Rafi, and S. Safavi-Naeini. Gain measurement of embedded on-chip antennas in mmw/thz range. *IEEE Transactions on Antennas and Propagation*, 60(5):2544–2549, May 2012.

[Niu05] Guofu Niu. Noise in SiGe HBT RF Technology: Physics, Modeling, and Circuit Implications. volume 93, pages 1583–1597, September 2005.

[NJB+16] I. Nasr, R. Jungmaier, A. Baheti, D. Noppeney, J. S. Bal, M. Wojnowski, E. Karagozler, H. Raja, J. Lien, I. Poupyrev, and S. Trotta. A Highly Integrated 60 GHz 6-Channel Transceiver With Antenna in Package for Smart Sensing and Short-Range Communications. *IEEE Journal of Solid-State Circuits*, 51(9):2066–2076, September 2016.

[NYP+08] S.T. Nicolson, K. H K Yau, S. Pruvost, V. Danelon, P. Chevalier, P. Garcia, A. Chantre, B. Sautreuil, and S.P. Voinigescu. A Low-Voltage SiGe BiCMOS 77-GHz Automotive Radar Chipset. volume 56, pages 1092–1104, May 2008.

[OHP12] E. Ojefors, B. Heinemann, and U.R. Pfeiffer. Subharmonic 220- and 320-GHz SiGe HBT Receiver Front-Ends. volume 60, pages 1397–1404, May 2012.

[OPCP10] E. Ojefors, F. Pourchon, P. Chevalier, and U.R. Pfeiffer. A 160-GHz low-noise downconverter in a SiGe HBT technology. In *European Microwave Conference (EuMC)*, pages 521–524, September 2010.

[OPLR09] E. Ojefors, U.R. Pfeiffer, A. Lisauskas, and H.G. Roskos. A 0.65 THz Focal-Plane Array in a Quarter-Micron CMOS Process Technology. volume 44, pages 1968–1976, July 2009.

[OR11] Y. C. Ou and G. M. Rebeiz. On-chip slot-ring and high-gain horn antennas for millimeter-wave wafer-scale silicon systems. *IEEE Transactions on Microwave Theory and Techniques*, 59(8):1963–1972, August 2011.

[OSMFSS09] M. Onabajo, J. Silva-Martinez, F. Fernandez, and E. Sanchez-Sinencio. An On-Chip Loopback Block for RF Transceiver Built-In Test. *IEEE Transactions on Circuits and Systems II: Express Briefs*, 56(6):444–448, June 2009.

[PCW+09] Piljae Park, L. Chen, Le Wang, S. Long, Hyun Kyu Yu, and C.P. Yue. On-wafer wireless testing and mismatch monitoring using RF transmitters with integrated antennas. In *IEEE Radio Frequency Integrated Circuits Symposium (RFIC)*, pages 505–508, June 2009.

[PDAO10] F. Poehl, F. Demmerle, J. Alt, and H. Obermeir. Production test challenges for highly integrated mobile phone SOCs- A case study. In *15th IEEE European Test Symposium (ETS)*, pages 17–22, May 2010.

[PFS+12] C. Pfeffer, R. Feger, C.M. Schmid, C. Wagner, and A. Stelzer. An IQ-modulator based heterodyne 77-GHz FMCW colocated MIMO radar system. In *IEEE International Microwave Symposium Digest (MTT),*, pages 1–3, 2012.

[POZ10] U.R. Pfeiffer, E. Ojefors, and Yan Zhao. A SiGe quadrature transmitter and receiver chipset for emerging high-frequency applications at 160GHz. In *IEEE International Solid-State Circuits Conference Digest of Technical Papers (ISSCC)*, pages 416–417, February 2010.

[PSL+07] Byung-Kwon Park, D. Samardzija, V.M. Lubecke, O. Boric-Lubecke, and T. Sizer. DC offset reduction in phase diversity heterodyne doppler radar system. In *IEEE Antennas and Propagation Society International Symposium,*, pages 3884–3887, June 2007.

[PWA+13] M. PourMousavi, M. Wojnowski, R. Agethen, R. Weigel, and A. Hagelauer. Antenna array in eWLB for 61 GHz FMCW radar. In *Asia-Pacific Microwave Conference Proceedings (APMC),*, pages 310–312, Nov 2013.

[Raz97] B. Razavi. Design considerations for direct-conversion receivers. *Circuits and Systems II: Analog and Digital Signal Processing, IEEE Transactions on*, 44(6):428–435, 1997.

[RHW+10] H. Rucker, B. Heinemann, W. Winkler, R. Barth, J. Borngraber, J. Drews, G.G. Fischer, A. Fox, T. Grabolla, U. Haak, D. Knoll, F. Korndorfer, A. Mai, S. Marschmeyer, P. Schley, D. Schmidt, J. Schmidt, M.A. Schubert, K. Schulz, B. Tillack, D. Wolansky, and Y.alib Yamamoto. A 0.13 μm SiGe BiCMOS Technology Featuring $f_T/f_{\max}$of 240/330 GHz and Gate Delays Below 3 ps. In *IEEE Journal of Solid-State Circuits*, volume 45, pages 1678–1686, September 2010.

[Rij96] J. J F Rijns. CMOS low-distortion high-frequency variable-gain amplifier. In *IEEE Journal of Solid-State Circuits*, volume 31, pages 1029–1034, Jul 1996.

[RKS06a] Jee-Youl Ryu, B.C. Kim, and I. Sylla. A new low-cost RF built-in self-test measurement for system-on-chip transceivers. *IEEE Transactions on Instrumentation and Measurement*, 55(2):381–388, 2006.

[RKS06b] Jee-Youl Ryu, B.C. Kim, and I. Sylla. A new low-cost rf built-in self-test measurement for system-on-chip transceivers. *IEEE Transactions on Instrumentation and Measurement,*, 55(2):381–388, April 2006.

[RV] David C. Rife and G.A. Vincent. Use of the discrete Fourier transform in the measurement of frequencies and levels of tones. *Belly Syst. Tech.*, 49(2):197–228.

[RV70] D. C. Rife and G. A. Vincent. Use of the discrete fourier transform in the measurement of frequencies and levels of tones. *The Bell System Technical Journal*, 49(2):197–228, February 1970.

[Sar13] I. Sarkas. *Cicuits ans System Design for mm-wave Radio and Radar Applications*. PhD thesis, University of Toronto, 2013.

[Sch16] Steffen Scherr. *FMCW-Radarsignalverarbeitung zur Entfernungsmessung mit hoher Genauigkeit*. PhD thesis, Karlsruher Institut of Technologie (KIT), 2016.

[SF92] S. Sampei and Kamilo Feher. Adaptive DC-offset compensation algorithm for burst mode operated direct conversion receivers. In *IEEE 42nd Vehicular Technology Conference, 1992,*, pages 93–96 vol.1, May 1992.

[SFCS12] Y. Sun, G.G. Fischer, and J. Christoph Scheytt. A Compact Linear 60-GHz PA With 29.2Avalanche Area in SiGe. volume 60, pages 2581–2589, August 2012.

[SGH+12] I. Sarkas, M.G. Girma, J. Hasch, T. Zwick, and S.P. Voinigescu. A Fundamental Frequency 143-152 GHz Radar Transceiver with Built-In Calibration and Self-Test. In *IEEE Compound Semiconductor Integrated Circuit Symposium (CSICS)*, pages 1–4, October 2012.

[SHBK07] Y. Sun, F. Herzel, J. Borngraber, and R. Kraemer. 60 GHz Receiver Building Blocks in SiGe BiCMOS. In *Topical Meeting on Silicon Monolithic Integrated Circuits in RF Systems*, pages 219–222, January 2007.

[SHBV12] I. Sarkas, J. Hasch, A. Balteanu, and S.P. Voinigescu. A Fundamental Frequency 120-GHz SiGe BiCMOS Distance Sensor With Integrated Antenna. volume 60, pages 795–812, March 2012.

[SHZ+09] A. Schicht, K. Huber, A. Ziroff, M. Willsch, and L. Schmidt. Absolute phase-based distance measurement for industrial monitoring systems. *IEEE Sensors Journal*, 9(9):1007–1013, September 2009.

[Sie58] K.M. Siegel. Bistatic radar and forward scattering. pages 286–290, 1958.

[Sko01] M. I. Skolnik. *Introduction to Radar Systems*. McGraw Hill, 2001.

[SLH+10] I. Sarkas, E. Laskin, J. Hasch, P. Chevalier, and S. P. Voinigescu. Second Generation transceivers for D-band Radar and Data Communication Applications. *IEEE MTT-S International Microwave Symposium Digest (MTT)*, pages 1328–1331, 2010.

[SLT+12] I. Sarkas, E. Laskin, A. Tomkins, A. Balteanu, E. Dacquay, L. Tarnow, S.P. Voinigescu, J. Hasch, P. Chevalier, and B. Sautreuil. Silicon-based radar and imaging sensors operating above 120 GHz. In *19th International Conference on Microwave Radar and Wireless Communications (MIKON)*, volume 1, pages 91–96, 2012.

[SM96] G. Strauss and W. S. Menzel. Millimeter-wave monolithic integrated circuit interconnects using electromagnetic field coupling. *IEEE Transactions on Components, Packaging, and Manufacturing Technology: Part B*, 19(2):278–282, May 1996.

[SMF+13] Y. Sun, M. Marinkovic, G. Fischer, W. Winkler, W. Debski, S. Beer, T. Zwick, M.G. Girma, J. Hasch, and C.J. Scheytt. A low-cost miniature 120GHz SiP FMCW/CW radar sensor with software linearization. In *IEEE International Solid-State Circuits Conference Digest of Technical Papers (ISSCC)*, pages 148–149, 2013.

[SPW+14] Y. Salamin, Jingnan Pan, Zhiyu Wang, Shuangtian Tang, Jian Wang, Changzhi Li, and Lixin Ran. Eliminating the Impacts of Flicker Noise and DC Offset in Zero-IF Architecture Pulse Compression Radars. *IEEE Transactions on Microwave Theory and Techniques*, 62(4):879–888, April 2014.

[SR05] R. Svitek and S. Raman. Dc offsets in direct-conversion receivers: characterization and implications. *IEEE Microwave Magazine*, 6(3):76–86, September 2005.

[SS11a] Y. Sun and C.J. Scheytt. A 122 GHz Sub-Harmonic Mixer With a Modified APDP Topology for IC Integration. volume 21, pages 679–681, December 2011.

[SS11b] Y. Sun and C.J. Scheytt. An integrated harmonic transmitter front-end for 122 GHz FMCW/CW radar sensor. In *European Microwave Integrated Circuits Conference (EuMIC)*, pages 97–100, 2011.

[SS12] Y. Sun and C.J. Scheytt. A low-phase-noise 61 GHz push-push VCO with divider chain and buffer in SiGe BiCMOS for 122 GHz ISM applications. In *IEEE Radio Frequency Integrated Circuits Symposium (RFIC)*, pages 79–82, June 2012.

[SSG+11] Y. Sun, K. Schmalz, S. Glisic, Ruoyu Wang, and C.J. Scheytt. An integrated differential 60 GHz sliding-IF receiver in SiGe BiCMOS. In *3rd International Conference on Computer Research and Development (ICCRD)*, volume 3, pages 297–300, March 2011.

[Sto92] A. G. Stove. Linear fmcw radar techniques. *IEE Proceedings F - Radar and Signal Processing*, 139(5):343–350, October 1992.

[SUC13] SUCCESS project website - http://www.success-project.eu. visited on February, 2013.

[SWB+10] K. Schmalz, W. Winkler, J. Borngraber, W. Debski, B. Heinemann, and J.C. Scheytt. A Subharmonic Receiver in SiGe Technology for 122 GHz Sensor Applications. volume 45, pages 1644–1656, September 2010.

[TDG+09] S. Trotta, B. Dehlink, A. Ghazinour, D. Morgan, and J. John. A 77 GHz 3.3V 4-channel transceiver in SiGe BiCMOS technology. In *IEEE Bipolar/BiCMOS Circuits and Technology Meeting, BCTM*, pages 186–189, October 2009.

[TEV13] K. Thurn, R. Ebelt, and M. Vossiek. Noise in homodyne fmcw radar systems and its effects on ranging precision. In *2013 IEEE MTT-S International Microwave Symposium Digest (MTT)*, pages 1–3, June 2013.

[Tuc54] D. G. Tucker. The history of the homodyne and synchrodyne. *Journal of the British Institution of Radio Engineers*, 14(4):143–154, April 1954.

[VGKB+07] A. Valdes-Garcia, W. Khalil, B. Bakkaloglu, J. Silva-Martinez, and E. Sanchez-Sinencio. Built-in Self Test of RF Transceiver SoCs: from Signal Chain to RF Synthesizers. In *IEEE Radio Frequency Integrated Circuits (RFIC) Symposium*, pages 335–338, June 2007.

[VGNL+10] A. Valdes-Garcia, S.T. Nicolson, Jie-Wei Lai, A. Natarajan, Ping-Yu Chen, S.K. Reynolds, J.-H.C. Zhan, D.G. Kam, Duixian Liu, and B. Floyd. A Fully Integrated 16-Element Phased-Array Transmitter in SiGe BiCMOS for 60-GHz Communications. volume 45, pages 2757–2773, December 2010.

[VHNM96] M. Vossiek, P. Heide, M. Nalezinski, and V. Magori. Novel FMCW radar system concept with adaptive compensation of phase errors. In *Microwave Conference, 1996. 26th European*, volume 1, pages 135–139, 1996.

[VLS+10] S.P. Voinigescu, E. Laskin, I. Sarkas, K. H K Yau, S. Shahramian, A. Hart, A. Tomkins, P. Chevalier, J. Hasch, P. Garcia, A. Chantre, and B. Sautreuil. Silicon D-band wireless transceivers and applications. In *Asia-Pacific Microwave Conference Proceedings (APMC)*, pages 1857–1864, 2010.

[VW03] Y Venot and W. Wiesbeck. 76.5 GHz radar sensor for contact-free distance measurement with micrometer accuracy. In *Proceedings of IEEE Sensors*, volume 1, pages 216–221 Vol.1, October 2003.

[WDH+09] W. Winkler, W. Debski, B. Heinemann, F. Korndorfer, H. Rucker, K. Schmalz, C. Scheytt, and B. Tillack. 122 GHz low-noise-amplifier in sige technology. In *ESSCIRC, 2009. ESSCIRC '09. Proceedings of*, pages 316–319, 2009.

[WDS+10] W. Winkler, W. Debski, K. Schmalz, J. Borngraeber, and Ch. Scheytt. LNA and Mixer for 122 GHz Receiver in SiGe Technology. In *proc. 40th European Microwave Conference (EuMC)*, pages 529–532, 2010.

[Wen05] J. Wenger. Automotive radar - status and perspectives. In *IEEE Compound Semiconductor Integrated Circuit Symposium, CSIC*, pages 4 pp.–, October 2005.

[Wil84] J. C. Wiltse. History of Millimeter and Submillimeter Waves. *IEEE Transactions on Microwave Theory and Techniques*, 32(9):1118–1127, September 1984.

[WLL10] D. Webster, J. Lopez, and D.Y.-C. Lie. How to do RF-BiST with virtually no extra circuits for RF-SoC products? In *Circuits and Systems (MWSCAS),53rd IEEE International Midwest Symposium on*, pages 469–472, 2010.

[WSJ06] C. Wagner, A. Stelzer, and H. Jager. PLL Architecture for 77-GHz FMCW Radar Systems with Highly-Linear Ultra-Wideband Frequency Sweeps. In *2006 IEEE MTT-S International Microwave Symposium Digest*, pages 399–402, June 2006.

[YE05] J.-S. Yoon and W.R. Eisenstadt. Embedded loopback test for RF ICs. *Instrumentation and Measurement, IEEE Transactions on*, 54(5):1715–1720, October 2005.

[ZdG05] A. Zjajo and J.P. de Gyvez. Evaluation of signature-based testing of RF/analog circuits. In *Test Symposium, 2005. European*, pages 62–67, May 2005.

[ZGA12] Chaoming Zhang, Ranjit Gharpurey, and JacobA. Abraham. Built-in Self Test of RF Subsystems with Integrated Detectors. *Journal of Electronic Testing*, 28(5):557–569, 2012.

[ZLG06] T. Zwick, D. Liu, and B. P. Gaucher. Broadband Planar Superstrate Antenna for Integrated Millimeterwave Transceivers. *IEEE Transactions on Antennas and Propagation*, 54(10):2790–2796, October 2006.

[ZPYM16] Y. Zeng, P. H. Pathak, Z. Yang, and P. Mohapatra. Poster Abstract: Human Tracking and Activity Monitoring Using 60 GHz mmWave. In *2016 15th ACM/IEEE International Conference on Information Processing in Sensor Networks (IPSN)*, pages 1–2, April 2016.

[ZSHvdW06] M. Zhao, J. D. Shea, S. C. Hagness, and D. W. van der Weide. Calibrated free-space microwave measurements with an ultrawideband reflectometer-antenna system. *IEEE Microwave and Wireless Components Letters*, 16(12):675–677, December 2006.

[ZXWH12] Bo Zhang, Yong-Zhong Xiong, Lei Wang, and Sanming Hu. A Switch-Based ASK Modulator for 10 Gbps 135 GHz Communication by 0.13 μm MOSFET. volume 22, pages 415–417, August 2012.

Forschungsberichte aus dem Institut für Höchstfrequenztechnik und Elektronik (IHE) der Universität Karlsruhe (TH)
(ISSN 0942-2935)

Herausgeber: Prof. Dr.-Ing. Dr. h.c. Dr.-Ing. E.h. mult. Werner Wiesbeck

Die Bände 1 (1992) bis 55 (2008) der Schriftenreihe können über das Institut Hochfrequenztechnik und Elektronik bestellt werden (https://www.ihe.kit.edu).

Fortführung als:

Karlsruher Forschungsberichte aus dem Institut für Hochfrequenztechnik und Elektronik
(ISSN 1868-4696)

Herausgeber: Prof. Dr.-Ing. Thomas Zwick

Band 55 Sandra Knörzer
Funkkanalmodellierung für OFDM-Kommunikationssysteme bei Hochgeschwindigkeitszügen (2009)
ISBN 978-3-86644-361-7

Band 56 Thomas Fügen
Richtungsaufgelöste Kanalmodellierung und Systemstudien für Mehrantennensysteme in urbanen Gebieten (2009)
ISBN 978-3-86644-420-1

Band 57 Elena Pancera
Strategies for Time Domain Characterization of UWB Components and Systems (2009)
ISBN 978-3-86644-417-1

Band 58 Jens Timmermann
Systemanalyse und Optimierung der Ultrabreitband-Übertragung (2010)
ISBN 978-3-86644-460-7

Band 59 Juan Pontes
Analysis and Design of Multiple Element Antennas for Urban Communication (2010)
ISBN 978-3-86644-513-0

Band 60 Andreas Lambrecht
True-Time-Delay Beamforming für ultrabreitbandige Systeme hoher Leistung (2010)
ISBN 978-3-86644-522-2

Band 61 Grzegorz Adamiuk
Methoden zur Realisierung von dual-orthogonal, linear polarisierten Antennen für die UWB-Technik (2010)
ISBN 978-3-86644-573-4

Band 62 Jutta Kühn
AlGaN/GaN-HEMT Power Amplifiers with Optimized Power-Added Efficiency for X-Band Applications (2011)
ISBN 978-3-86644-615-1

Band 63 Małgorzata Janson
Hybride Funkkanalmodellierung für ultrabreitbandige MIMO-Systeme (2011)
ISBN 978-3-86644-639-7

Band 64 Mario Pauli
Dekontaminierung verseuchter Böden durch Mikrowellenheizung (2011)
ISBN 978-3-86644-696-0

Band 65 Thorsten Kayser
Feldtheoretische Modellierung der Materialprozessierung mit Mikrowellen im Durchlaufbetrieb (2011)
ISBN 978-3-86644-719-6

Band 66 Christian Andreas Sturm
Gemeinsame Realisierung von Radar-Sensorik und Funkkommunikation mit OFDM-Signalen (2012)
ISBN 978-3-86644-879-7

Die Bände sind unter www.ksp.kit.edu als PDF frei verfügbar oder als Druckausgabe bestellbar.

Band 67 Huaming Wu
Motion Compensation for Near-Range Synthetic Aperture Radar Applications (2012)
ISBN 978-3-86644-906-0

Band 68 Friederike Brendel
Millimeter-Wave Radio-over-Fiber Links based on Mode-Locked Laser Diodes (2013)
ISBN 978-3-86644-986-2

Band 69 Lars Reichardt
Methodik für den Entwurf von kapazitätsoptimierten Mehrantennensystemen am Fahrzeug (2013)
ISBN 978-3-7315-0047-6

Band 70 Stefan Beer
Methoden und Techniken zur Integration von 122 GHz Antennen in miniaturisierte Radarsensoren (2013)
ISBN 978-3-7315-0051-3

Band 71 Łukasz Zwirełło
Realization Limits of Impulse-Radio UWB Indoor Localization Systems (2013)
ISBN 978-3-7315-0114-5

Band 72 Xuyang Li
Body Matched Antennas for Microwave Medical Applications (2014)
ISBN 978-3-7315-0147-3

Band 73 Sebastian Diebold
Transistor- und Leitungsmodellierung zum Entwurf von monolithisch integrierten Leistungsverstärkern für den hohen Millimeterwellen-Frequenzbereich (2014)
ISBN 978-3-7315-0161-9

Band 74 Christian Rusch
Integrierte, planare Leckwellenantennen für 3D-Millimeterwellen-Radarsysteme basierend auf dem holografischen Prinzip (2014)
ISBN 978-3-7315-0234-0

Band 75 Marlene Harter
Dreidimensional bildgebendes Radarsystem mit digitaler Strahlformung für industrielle Anwendungen (2014)
ISBN 978-3-7315-0249-4

Band 76 Michael A. Baldauf
Abhängigkeit der Exposition von der Zellgröße beim Mobilfunk unter Gewährleistung der Versorgung (2015)
ISBN 978-3-7315-0308-8

Band 77 Alicja Ossowska
Highly Resolved Synthetic Aperture Radar with Beam Steering (2015)
ISBN 978-3-7315-0315-6

Band 78 Małgorzata Dominika Brzeska
RF Modelling and Characterization of Tyre Pressure Sensors and Vehicle Access Systems (2015)
ISBN 978-3-7315-0348-4

Band 79 Ulrich Lewark
Aktive Frequenzvervielfacher zur Signalerzeugung im Millimeter- und Submillimeterwellen Frequenzbereich (2015)
ISBN 978-3-7315-0354-5

Band 80 Kai-Philipp Walter Pahl
Distributed Transformers for Broadband Monolithic Millimeter-Wave Integrated Power Amplifiers (2015)
ISBN 978-3-7315-0409-2

Band 81 Serdal Ayhan
Hochgenaue radarbasierte Abstandsmessung mit geführter Wellenausbreitung (2016)
ISBN 978-3-7315-0433-7

Band 82 Yoke Leen Sit
MIMO OFDM Radar-Communication System with Mutual Interference Cancellation (2017)
ISBN 978-3-7315-0599-0

Band 83 Steffen Scherr
FMCW-Radarsignalverarbeitung zur Entfernungsmessung mit hoher Genauigkeit (2017)
ISBN 978-3-7315-0607-2

Band 84 Tom Schipper
Modellbasierte Analyse des Interferenzverhaltens von Kfz-Radaren (2017)
ISBN 978-3-7315-0639-3

Band 85 Malyhe Jalilvand
Application-Specific Broadband Antennas for Microwave Medical Imaging (2017)
ISBN 978-3-7315-0664-5

Band 86 Benjamin Göttel
Millimeterwellen On-Chip Antennensysteme für die Integration in SoC Applikationen (2017)
ISBN 978-3-7315-0667-6

Band 87 Christian Arnold
Im Orbit einstellbare Ausgangsfilter und -multiplexer (2017)
ISBN 978-3-7315-0722-2

Band 88 Tobias Mahler
Synthese kapazitätsoptimierter Antennensysteme mit messtechnischer Verifikation (2018)
ISBN 978-3-7315-0737-6

Band 89 Daniel Müller
RF Probe-Induced On-Wafer Measurement Errors in the Millimeter-Wave Frequency Range (2018)
ISBN 978-3-7315-0822-9

Band 90 Tristan Visentin
Polarimetric Radar for Automotive Applications (2019)
ISBN 978-3-7315-0888-5

Band 91 Christian von Vangerow
Entwurf und Modellierung von Breitbandverstärkern mit variablem Gewinn in SiGe BiCMOS Technologien (2019)
ISBN 978-3-7315-0910-3

Band 92 Mekdes Girma
Concepts for Short Range Millimeter-wave Miniaturized Radar Systems with Built-in Self-Test (2019)
ISBN 978-3-7315-0938-7